WARREN W. TRYON, Ph.D.
DEPARTMENT OF PSYCHOLOGY
FORDHAM UNIVERSITY
BRONX, NEW YORK 10458

# BASIC STATISTICS

# BASIC

McGraw-Hill Book Company

# STATISTICS

**DAVID BLACKWELL**
Department of Statistics
University of California, Berkeley

New York   San Francisco   St. Louis   Toronto   London   Sydney

**BASIC STATISTICS**

Copyright © 1969 by McGraw-Hill, Inc. All rights reserved. No part of this publication may be reproduced, stored in a retrieval system, or transmitted, in any form or by any means, electronic, mechanical, photocopying, recording, or otherwise, without the prior written permission of the publisher.

Printed in the United States of America.

Library of Congress catalog card number: 69-16250

1234567890 MAMM 754321069     **05531**

# PREFACE

This book indicates the content of a lower-division basic statistics course I have taught several times at Berkeley. The students come from all departments of the university, and many of them have forgotten high-school algebra. The mathematical level of the course is modest: Any student who can do arithmetic, substitute in simple formulas, plot points, and draw a smooth curve through plotted points is ready for the course. But he must be prepared to think seriously about frivolous examples, as balls in urns are used to illustrate practically every idea introduced. The approach is intuitive, informal, concrete, decision-theoretic, and Bayesian.

David Blackwell

# CONTENTS

| | | |
|---|---|---|
| 1 | PROBABILITY | 1 |
| 2 | VARIABLES | 12 |
| 3 | DENSITIES | 19 |
| 4 | MEAN | 26 |
| 5 | VARIANCE | 36 |
| 6 | WORTH OF A PREDICTOR | 46 |

| 7 | CORRELATION | 51 |
|---|---|---|
| 8 | MULTIPLE AND PARTIAL CORRELATION | 59 |
| 9 | INDEPENDENCE | 64 |
| 10 | BINOMIAL DISTRIBUTION | 72 |
| 11 | NORMAL APPROXIMATION | 80 |
| 12 | INFERENCE | 88 |
| 13 | INFERENCE ABOUT PROPORTIONS (I) | 96 |
| 14 | INFERENCE ABOUT PROPORTIONS (II) | 102 |
| 15 | INDEPENDENT PROPORTIONS | 108 |
| 16 | CHI SQUARE | 114 |
| | ANSWERS | 121 |
| | APPENDIX | 139 |
| | INDEX | 141 |

# 1 PROBABILITY

If an urn contains five balls and we select one of these balls in such a way that each of the five balls has the same chance of being selected, we say that the ball was *selected at random*, and that the *probability* or *chance* that any particular ball was selected is $\frac{1}{5}$. If three of the five balls are red, the probability that the selected ball is red is $\frac{3}{5}$. In general, when an element is selected at random from a finite set $S$, so that each element of $S$ is equally likely to be selected, the probability that the selected element is in a given subset $A$ of $S$ is defined as the ratio of the number of elements in $A$ to the number of elements in $S$:

For random selection from $S$,
$$P(A) = \frac{\text{number of elements in } A}{\text{number of elements in } S}$$

**EXAMPLE 1.1** One of the following hundred numbers is selected at random:

$$\begin{array}{cccc} 00 & 01 & \cdots & 09 \\ 10 & 11 & \cdots & 19 \\ \multicolumn{4}{c}{\cdots\cdots\cdots\cdots} \\ 90 & 91 & \cdots & 99 \end{array}$$

Some events and their probabilities are as follows:

|     | Events | Probabilities |
| --- | --- | --- |
| (a) | The first digit is 0 | .1 |
| (b) | The two digits are equal | .1 |
| (c) | The two digits are unequal | .9 |
| (d) | The first digit is larger than the second | .45 |
| (e) | The first digit is at least as large as the second | .55 |
| (f) | The second digit is 1 | .1 |
| (g) | The sum of the two digits is 5 | .06 |
| (h) | The sum of the two digits is 9 | .1 |
| (i) | Neither digit exceeds 3 | .16 |
| (j) | Both digits exceed 3 | .36 |
| (k) | Only one of the digits exceeds 3 | .48 |
| (l) | The first digit exceeds 3 and the second does not | .24 |

In general, the probability of any event is a number between 0 and 1, measuring how likely the event is to occur. Probabilities satisfy these rules:

1. $P(A) = 0$ if $A$ is *impossible*, that is, cannot occur
2. $P(A) = 1$ if $A$ is *certain*, that is, must occur
3. $P(A \text{ or } B) = P(A) + P(B)$ if $A$ and $B$ are *mutually exclusive*, that is, cannot both occur
4. $P(\text{not } A) = 1 - P(A)$

**EXAMPLE 1.2** A point is selected at random in a square $S$ of side 1, so that the probability that the selected point is in a given subset $A$ of $S$ equals the area of $A$:

$P(A)$ = area of $A$

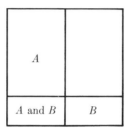

**FIGURE 1.1**

As diagramed in Figure 1.1, with $A$ the left half of $S$ and $B$ the lower quarter of $S$:

$P(A) = \frac{1}{2}$   $P(B) = \frac{1}{4}$   $P(A \text{ and } B) = \frac{1}{8}$

since the points in $A$ and $B$ are the points in the lower left rectangle, and $P(A \text{ or } B) = \frac{5}{8}$, since all points not in the upper right rectangle are in at least one of $A$ and $B$, that is, are in $A$ or $B$. With $C = \text{not}(A \text{ or } B)$, the upper right rectangle:

$P(C) = \frac{3}{8}$   $P(A \text{ or } C) = P(A) + P(C) = \frac{1}{2} + \frac{3}{8} = \frac{7}{8}$

**EXAMPLE 1.3** Box 1 contains five balls, labeled 1, 2, 3, 4, 5, and box 2 contains five balls, labeled 6, 7, 8, 9, 10. One of the two boxes is selected at random, then a ball is selected at random from the selected box. You win a prize if the number on the selected ball is divisible by 3, that is, if it is one of the numbers 3, 6, or 9. Since the selection of any one of the 10 balls is equally likely, $P(\text{prize}) = \frac{3}{10} = .3$. Suppose box 1 is selected. Now what is the probability that you win the prize? Since occurrence of any one of the five numbers 1, 2, 3, 4, and 5 is now equally likely, and only one of them wins the prize, the probability that you win the prize given that box 1 is selected is $\frac{1}{5} = .2$.

We write this

$$P(\text{prize}|\text{box 1}) = .2$$

Similarly

$$P(\text{prize}|\text{box 2}) = .4$$

In general, for any two events $A$ and $B$, the probability of $B$ given that $A$ occurs is denoted by $P(B|A)$, and satisfies the formula:

$$P(B|A) = \frac{P(A \text{ and } B)}{P(A)}$$

In our example, with $A = $ box 1 and $B = $ prize, $A$ and $B$ have the single element 3, so $P(A \text{ and } B) = .1$. Since $P(A) = .5$, the formula gives $P(B|A) = .1/.5 = .2$, agreeing with what we found directly.

The formula for $P(B|A)$ can also be written

$$P(A \text{ and } B) = P(A)P(B|A)$$

and, in this form, it can be extended to more than two events:

**Multiplication Rule**

$P(A \text{ and } B \text{ and } C \text{ and } \ldots) = P(A)P(B|A)P(C|A \text{ and } B) \ldots$ The probability that several events all occur is the probability that one occurs, times the probability that a second occurs given that the first occurs, times the probability that a third occurs given that the first two both occur, and so on.

For instance, for three events we get

$$P(A \text{ and } B \text{ and } C) = P(A \text{ and } B)P(C|A \text{ and } B)$$
$$= P(A)P(B|A)P(C|A \text{ and } B)$$

EXAMPLE 1.4    The urn shown in Figure 1.2 contains two black balls and three white ones. Two balls are drawn successively at random from the urn. Find the probability of each of the four outcomes $BB$, $BW$, $WB$, and $WW$:

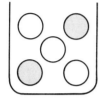

FIGURE 1.2

(a) If the first ball is not replaced in the urn before the second ball is drawn (drawing *without* replacement)
(b) If the first ball is replaced (drawing *with* replacement)

Denote by $B_1$ the event that the first ball is black, and so forth. In case (a), as diagramed in Figure 1.3,

$$P(BB) = P(B_1 \text{ and } B_2) = P(B_1)P(B_2|B_1) = \tfrac{2}{5}(\tfrac{1}{4}) = .1$$
$$P(BW) = P(B_1)P(W_2|B_1) = \tfrac{2}{5}(\tfrac{3}{4}) = .3$$
$$P(WB) = P(W_1)P(B_2|W_1) = \tfrac{3}{5}(\tfrac{2}{4}) = .3$$
$$P(WW) = P(W_1)P(W_2|W_1) = \tfrac{3}{5}(\tfrac{2}{4}) = .3$$

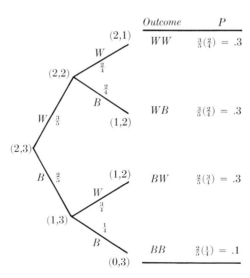

FIGURE 1.3

**BASIC STATISTICS**

In case (b)

$P(BB) = P(B_1)P(B_2|B_1) = \frac{2}{5}(\frac{2}{5}) = .16$
$P(BW) = P(B_1)P(W_2|B_1) = \frac{2}{5}(\frac{3}{5}) = .24$
$P(WB) = P(W_1)P(B_2|W_1) = \frac{3}{5}(\frac{2}{5}) = .24$
$P(WW) = P(W_1)P(W_2|W_1) = \frac{3}{5}(\frac{3}{5}) = .36$

**EXAMPLE 1.5** A woman sometimes buys brand $A$, and sometimes brand $B$. If her last purchase was satisfactory, she buys the same brand; if not, she changes. If a given purchase of brand $A$ has probability .7 of being satisfactory, while a given purchase of brand $B$ has probability .8, what is the probability that her third purchase will be brand $A$, if she tosses a coin to decide the first purchase? The diagram in Figure 1.4 shows that

$P(\text{third purchase is } A) = .245 + .030 + .070 + .080$
$= .425$

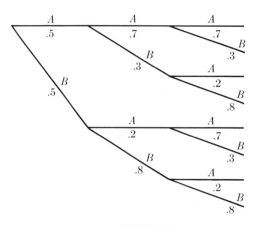

**FIGURE 1.4**

**EXAMPLE 1.6**
**Polya Urn Scheme**

A club consists of two kinds of people: $R$'s and $D$'s. Each year, one member of the club is selected at random, and allowed to choose a new member. An $R$ will always choose an $R$, and a $D$ will always choose a $D$. If the club originally consists of one $R$ and one $D$, what is the chance that, after three selections have been made, the club will consist of three $R$'s and two $D$'s? The diagram in Figure 1.5 shows that

$$P(3,2) = \tfrac{1}{12} + \tfrac{1}{12} + \tfrac{1}{12} = \tfrac{1}{4}$$

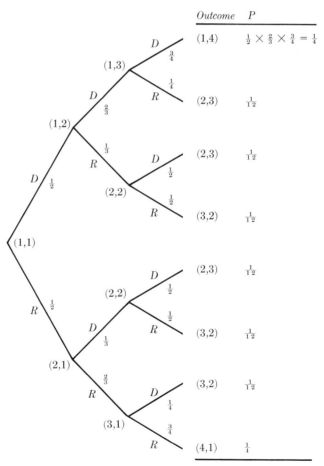

**FIGURE 1.5**

**EXAMPLE 1.7** Only one person in 10 in a certain population has hylosis. Of those with hylosis, 80% will react positively to the $X$ test, while only 30% of those without hylosis will react positively. One person is selected at random from the population and given the $X$ test. What is the probability that he has hylosis if he reacts positively? Negatively? The diagram in Figure 1.6 shows that

$$P(H|+) = \frac{P(H \text{ and } +)}{P(+)} = \frac{.08}{.08 + .27} = \frac{8}{35} = .23$$

$$P(H|-) = \frac{P(H \text{ and } -)}{P(-)} = \frac{.02}{.02 + .63} = \frac{2}{65} = .031$$

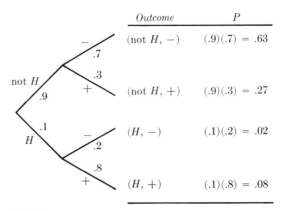

**FIGURE 1.6**

**EXAMPLE 1.8** Every day, one of five people is selected at random. What is the probability that, on the first three days:

(a) The same person is chosen every day?
(b) Three different people are chosen?

In Figure 1.7, the numbers at the vertices indicate the number of different people who have been chosen. Thus

$P(\text{same person every day}) = .04$
$P(\text{three different people}) = .48$

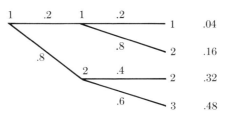

**FIGURE 1.7**

## PROBLEMS
## Set 1

Draw diagrams whenever possible.

1. One of the four words in the sentence THE URN IS COLD is selected at random. Find the probability that the number of letters in the selected word is: (a) 2, (b) 3, (c) 4.

2. One of the 12 letters in the sentence THE URN IS COLD is selected at random. Find the probability that the number of letters in the word containing the selected letter is: (a) 2, (b) 3, (c) 4.

3. A fair coin is tossed until it shows heads or until four tails are obtained. Find the probability of each of the five outcomes $H, TH, TTH, TTTH, TTTT$.

4. A committee is made up of four males and two females. Two members are selected successively at random without replacement. (a) Find the probability of each of the four outcomes $MM, MF, FM, FF$. (b) What is the most probable outcome? (c) What is the most probable number of males?

5. An urn contains two balls labeled $A$, one labeled $B$, and one labeled $C$. The four balls are drawn successively at random, and so form a four-letter "word." What is the probability that the selected word begins with: (a) $A$, (b) $B$, (c) $C$, (d) $AA$, (e) $AB$, (f) $ABA$?

6. An urn contains two white balls and two black ones. Balls are drawn successively at random from the urn, without replacement. (a) What is the probability that the first black ball appears on the first draw? The second

draw? The third? (b) What is the probability that the second black ball appears on the second draw? The third draw? The fourth?

7. How would you go about selecting one of five houses at random? Could you do it with a fair die? With a fair coin?

8. Room $A$ contains three people and room $B$ contains two. One of the rooms is selected at random; then a person is selected at random from the selected room and given a prize. What is your chance of getting the prize if you are in room $A$? In room $B$?

9. You give a friend a letter to mail. The probability that he forgets to mail it is .1, the probability that the post office fails to deliver it, given that it was mailed, is .1, and the probability that the addressee fails to get the letter, given that it was delivered, is .1. Given that the addressee fails to get the letter, what is the probability that your friend forgot to mail it?

10. Three people are selected at random with replacement from a population of 100. What is the probability that three different people are chosen?

11. The proportion of females in a certain population is $p$. Five people are selected at random with replacement. What is the probability of the outcome $MFFFF$: (a) If $p = .5$? (b) If $p = .2$?

12. College 1 contains 50% females, and college 2 contains 20% females. One of the two colleges is selected at random, then five students are selected at random from the selected college. What is the probability that college 1 was selected if the sample is $MFFFF$?

13. A part has probability .9 of working. A component consists of three parts connected in series, so that the component works only if all three parts work. A machine consists of two components connected in parallel, so that the machine works if either of its components works. What is the probability that the machine works?

14. Each time a fair coin falls heads $B$ pays $A$ $1, and each time it falls tails $A$ pays $B$ $1. They start with $2 each, and continue playing until one of them is broke. (a) What is the probability that the game terminates after two tosses? After four tosses? (b) What is the probability that $A$ wins, given that the first toss is heads?

15. A man starts in the center of town and tosses a fair coin. If it shows heads, he walks one block east; if tails, one block north. Then he tosses the coin again and uses the same rule. Considering the center of town as the origin $(0,0)$, what is the probability that he is at $(3,1)$ after four tosses?

# 2 VARIABLES

Any rule that associates with each outcome of an experiment a corresponding number is called a *variable*. The number that a variable associates with a particular outcome is called the *value* of the variable for that outcome. A list of the values of the variable, together with their corresponding probabilities, is called the *distribution* of the variable.

EXAMPLE 2.1   Two students are selected at random (with replacement) from a college in which 60% of the students are male. The number $X$ of males in the sample is a variable.

From the diagram given in Figure 2.1, we can obtain the following distribution of $X$:

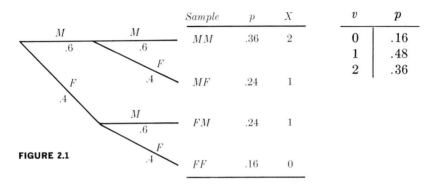

FIGURE 2.1

| Sample | $p$ | $X$ |
|---|---|---|
| $MM$ | .36 | 2 |
| $MF$ | .24 | 1 |
| $FM$ | .24 | 1 |
| $FF$ | .16 | 0 |

| $v$ | $p$ |
|---|---|
| 0 | .16 |
| 1 | .48 |
| 2 | .36 |

The distribution of a variable can be represented by a picture, called a *histogram*. Figure 2.2 shows the histogram for $X$. It consists of nonoverlapping rectangles with

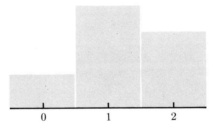

FIGURE 2.2

equal bases, one centered at each value of the variable, with height proportional to the probability that the variable has that value. The heights of our rectangles are proportional to 16, 48, and 36.

For any two variables $X$ and $Y$, we denote by $X + Y$ the variable whose value for each outcome is the value of $X$ for that outcome plus the value of $Y$ for that outcome. The variables $X - Y$, $XY$, $X/Y$, $X^2$, $3X + 2Y - 7$, and so on are defined analogously.

**EXAMPLE 2.2** One of the five words in the following sentence is selected at random:

THERE IS NO LARGEST INTEGER

Find the distributions of:

$V$, the number of vowels in the selected word
$C$, the number of consonants
$V + C$
$C - V - 1$
$(V - 2)^2$

| Element | $V$ | $C$ | $V+C$ | $C-V-1$ | $(V-2)^2$ |
|---|---|---|---|---|---|
| THERE | 2 | 3 | 5 | 0 | 0 |
| IS | 1 | 1 | 2 | $-1$ | 1 |
| NO | 1 | 1 | 2 | $-1$ | 1 |
| LARGEST | 2 | 5 | 7 | 2 | 0 |
| INTEGER | 3 | 4 | 7 | 0 | 1 |

The distributions are:

$V$:

| $v$ | $p$ |
|---|---|
| 1 | .4 |
| 2 | .4 |
| 3 | .2 |

$C$:

| $v$ | $p$ |
|---|---|
| 1 | .4 |
| 3 | .2 |
| 4 | .2 |
| 5 | .2 |

$V + C$:

| $v$ | $p$ |
|---|---|
| 2 | .4 |
| 5 | .2 |
| 7 | .4 |

$C - V - 1$:

| $v$ | $p$ |
|---|---|
| $-1$ | .4 |
| 0 | .4 |
| 2 | .2 |

$(V - 2)^2$:

| $v$ | $p$ |
|---|---|
| 0 | .4 |
| 1 | .6 |

# VARIABLES

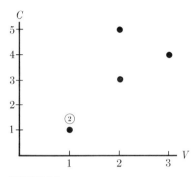

**FIGURE 2.3**

From the separate distributions of $V$ and $C$ we learn nothing about their joint behavior, such as whether large values of $V$ tend to be associated with large values of $C$. Such information is given by a picture called a *scatter diagram*.

Figure 2.3 shows the scatter diagram for $V$ and $C$. We plot all pairs of values of the variables, and label each point with a number proportional to its probability; unlabeled points are understood to have weight 1. The points tend to rise as we go to the right: large values of $V$ tend to be associated with large values of $C$.

**DISCUSSION PROBLEM 2.1** For each of the scatter diagrams of Figure 2.4, find the distributions and draw the histograms for $X$, $Y$, $X + Y$, $X - Y$, and find $P(X > Y)$, $P(Y = 2X + 1)$.

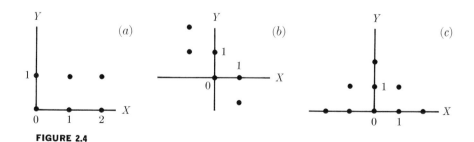

**FIGURE 2.4**

## PROBLEMS Set 2

1. An urn contains three balls labeled 0, 1, and 2 respectively. Two balls are drawn at random without replacement. The number on the first ball drawn is $X$ and the number on the second ball drawn is $Y$. (a) Find the distribution and draw the histogram for each of these variables: $X$, $Y$, $X + Y$, $X - Y$, $|X - Y|$. (b) Draw the scatter diagram for $X$ and $Y$.

2. Repeat Problem 1 but assume the balls are drawn with replacement.

3. Repeated drawings with replacement are made from an urn in which the proportion of black balls is .8. You are paid \$1 every time a black ball is drawn, but paid nothing otherwise. $X_i$ denotes your income from the $i$th draw, and $S_i = X_1 + \cdots + X_i$ denotes your total income from the first $i$ draws. (a) Find the distribution and draw the histogram for each of these variables: $X_1$, $S_2$, $S_3$, $S_3 - S_2$, $X_1 - X_2$. (b) Draw the scatter diagram for the two variables $S_2$ and $S_3$.

4. Draw the histogram for the variable $S_3$ in Problem 3, for each of these proportions of black balls in the urn: .6, .5, .1, 0, 1.

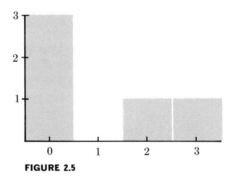

**FIGURE 2.5**

5. Figure 2.5 shows the histogram for a variable $X$. Draw the histogram for each of these variables: $X + 2$, $2X$, $X - 1$, $X^2$, $(X - 1)^2$, $X/(X + 1)$.

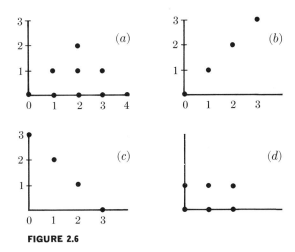
**FIGURE 2.6**

6. For each of the scatter diagrams shown in Figure 2.6, find the distributions of $X$, $Y$, and $X + Y$, draw their histograms, and draw the scatter diagram for the variables $X$ and $Y - X$.

7. An urn contains four balls, labeled 0, 1, 2, and 3. Repeated drawings with replacement are made, and $X_n$ denotes the largest number observed in the first $n$ draws. Find $P(X_2 = 0)$, $P(X_2 \leq 1)$, $P(X_2 \leq 2)$. Draw the histogram for $X_2$. Draw the histogram for $X_3$.

8. An urn contains five balls; three of them are labeled 0, one is labeled 1, and one is labeled 2. The balls are drawn successively at random, without replacement.

(a) List the possible outcomes of the first two draws, and their probabilities.

(b) List the possible outcomes of the first three draws, and their probabilities.

(c) Find the distribution of $X$, the sum of the labels of the first two balls drawn from the list in (a), and from the list in (b).

(d) Find the distribution of $(X - 1)^2$ from the list in (a), from the list in (b), and directly from the distribution of $X$.

9. A point is selected at random in the square with vertices (0,0), (1,0), (1,1), and (0,1). Denote by $X$ and $Y$ the $x,y$ coordinates of the selected point. (a) Find $P(X + Y \leq t)$ for each of these values of $t$: 0, .5, 1, 1.5, 2. (b) Graph $P(X + Y \leq t)$ as a function of $t$ for $0 \leq t \leq 2$.

10. There are three people on a committee. Each year one of them, selected at random, changes his mind about capital punishment. Initially all favor it. If $X_n$ denotes the number of people who favor capital punishment in year $n$ (so $X_1 = 3$), find the distributions and draw the histograms for $X_2$, $X_3$, and $X_4$. Draw the scatter diagram for $X_3$ and $X_4$.

11. A man has a large number of watches to repair; 40% of them will require one hour of work, 20% two hours, and 40% three hours. He selects two watches at random and repairs them. Find the distribution of $X$, the time it takes him to repair the two selected watches.

# 3 DENSITIES

Suppose the distribution of the variable $H$, height to the nearest inch in a certain population, is as follows:

| $H$ | Proportion with that value of $H$ |
|---|---|
| 63 | .01 |
| 64 | .03 |
| 65 | .07 |
| 66 | .12 |
| 67 | .17 |
| 68 | .20 |
| 69 | .17 |
| 70 | .12 |
| 71 | .07 |
| 72 | .03 |
| 73 | .01 |

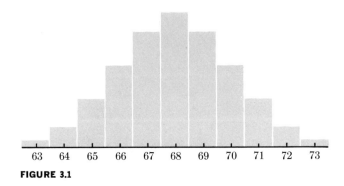
**FIGURE 3.1**

Figure 3.1 shows a histogram for $H$. If we had measured height to the nearest hundredth of an inch, we would have 1100 values for $H$, the histogram would have 1100 rectangles (thin ones if we used the same scale for $H$) and, if adjacent rectangles had about the same height, the histogram would be very much like the smooth curve of Figure 3.2.

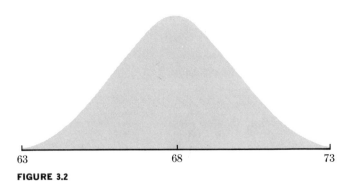
**FIGURE 3.2**

Any curve that is the graph of a nonnegative function and includes a finite positive area between itself and the $x$ axis can be considered as an approximation to the histogram of a variable $X$ with a large number of values. Such a curve is called a *density*. For any two numbers $a$ and $b$, the probability that the value of $X$ is between $a$

# DENSITIES

and $b$ is proportional to the area under the density between $a$ and $b$:

$$P(a \leq X \leq b) = \frac{\text{area under density between } a \text{ and } b}{\text{total area under density}}$$

**EXAMPLE 3.1**  Figure 3.3 shows the density for a variable $X$. Find $P(11 \leq X \leq 13)$.

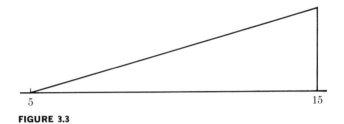

**FIGURE 3.3**

In Figure 3.4, $P(11 \leq X \leq 13)$ is the ratio of the shaded area to the area of the triangle $ABC$. Say the height of the triangle at 15 is 1 (any positive number will do). Then the area of the shaded trapezoidal region is $\frac{.6 + .8}{2}(2) = 1.4$ and the area of the triangle is $\frac{1}{2} \times 10 \times 1 = 5$, thus

$$P(11 \leq X \leq 13) = \frac{1.4}{5} = .28$$

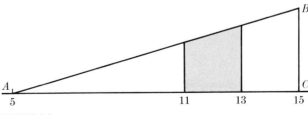

**FIGURE 3.4**

For our $X$, any value between 5 and 15 is possible. If we decide to measure $X$ approximately, say to the nearest even integer, the approximating variable $X^*$ would have only 6, 8, 10, 12, and 14 as values:

Whenever $X$ is between 5 and 7, $X^* = 6$

Whenever $X$ is between 7 and 9, $X^* = 8$

. . . . . . . . . . . . . . . . . . . . . . . . .

Whenever $X$ is between 13 and 15, $X^* = 14$

We call $X^*$ the *approximation* to $X$ with values 6, 8, 10, 12, and 14.

The distribution of $X^*$ is as follows:

| $v$ | $p$ |
|---|---|
| 6 | .04 |
| 8 | .12 |
| 10 | .20 |
| 12 | .28 |
| 14 | .36 |

Figure 3.5 shows the histogram for $X^*$, together with the density of $X$.

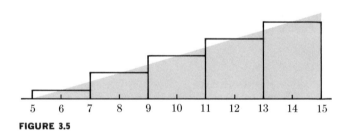

**FIGURE 3.5**

**EXAMPLE 3.2** Figure 3.6 shows the density of a variable $X$. What is the distribution of the approximating variable $X^*$ with 1, 3, 5, 7, and 9 as values?

We must find the areas of the five regions separated by the vertical lines. Taking the vertical scale 10, 20, 30 as shown (we could take 1, 2, 3 or 2, 4, 6), the heights

# DENSITIES

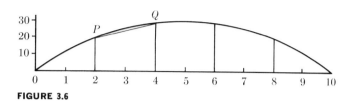

**FIGURE 3.6**

of the vertical lines are (by measurement) 21, 31, 31, and 21. To find the area between 2 and 4, we ignore the small shaded area between the line $PQ$ and the density and find the area of the trapezoidal region below $PQ$ to be $\frac{21+31}{2}$ (2) = 52. Find the other four areas similarly:

Area under density between 0 and  2 =  21
Area under density between 2 and  4 =  52
Area under density between 4 and  6 =  62
Area under density between 6 and  8 =  52
Area under density between 8 and 10 =  21
Total area under density                 = 208

Thus

$P(0 \leq X \leq 2) = P(X^* = 1) = \frac{21}{208} = .10$
$P(2 \leq X \leq 4) = P(X^* = 3) = \frac{52}{208} = .25$
$P(4 \leq X \leq 6) = P(X^* = 5) = \frac{62}{208} = .30$
$P(6 \leq X \leq 8) = P(X^* = 7) = \frac{52}{208} = .25$
$P(8 \leq X \leq 10) = P(X^* = 9) = \frac{21}{208} = .10$

We may always choose the vertical scale so that the total area under a density is 1. Then the area associated with each set of values of the variable becomes the probability that the value of the variable is in that set. For example, if we take the height $CB$ in Figure 3.4 to be .2, the area of triangle $ABC$ becomes 1, and the shaded area is $\frac{.12 + .16}{2}$ (2) = .28, which we calculated as $P(11 \leq X \leq 13)$.

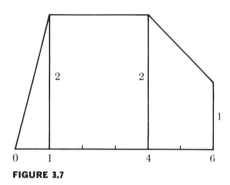

**FIGURE 3.7**

**PROBLEMS Set 3**

1. Figure 3.7 shows the density for a variable $X$. Check that the total area under the density is 10. Find:

   (a) $P(0 \leq X \leq 1)$
   (b) $P(1 \leq X \leq 4)$
   (c) $P(4 \leq X \leq 6)$

   What is the distribution of the approximating variable $X^*$ with 1, 3, and 5 as values? Graph $P(X \leq t)$ as a function of $t$ for $0 \leq t \leq 2$. Graph $P(X^* \leq t)$ for $0 \leq t \leq 6$.

2. Draw the graph of each of the following functions, and consider it as the density of a variable $X$:

   (a) $f(t) = \begin{cases} 1 & 0 \leq t \leq 8 \\ 0 & \text{otherwise} \end{cases}$ uniform or rectangular density on $0 \leq t \leq 8$

   (b) $f(t) = \begin{cases} t & 0 \leq t \leq 8 \\ 0 & \text{otherwise} \end{cases}$

   (c) $f(t) = \begin{cases} 8 - t & 0 \leq t \leq 8 \\ 0 & \text{otherwise} \end{cases}$

   (d) $f(t) = \begin{cases} t(8 - t) & 0 \leq t \leq 8 \\ 0 & \text{otherwise} \end{cases}$

   Find the distribution of the approximating $X^*$ with values 1, 3, 5, and 7.

3. Draw a semicircle with center at (4,0), radius 4, and diameter on the $x$ axis. Consider it as a density and find,

approximately, the distribution of the approximating variable with values 1, 3, 5, and 7.

4. For a certain variable $X$:

$$P(X \leq t) = \begin{cases} 0 & \text{for } t \leq 0 \\ \dfrac{t}{6} & \text{for } 0 \leq t \leq 6 \\ 1 & \text{for } t \geq 6 \end{cases}$$

(a) Graph $P(X \leq t)$ as a function of $t$.
(b) Find $P(X \leq 2)$, $P(X \leq 4)$, and $P(2 \leq X \geq 4)$.
(c) What is the distribution of $X^*$ with values 1, 3, and 5? Draw the histogram for $X^*$. What is the density for $X$?

5. Repeat Problem 4 with:

$$P(X \leq t) = \frac{t^2}{36} \text{ for } 0 \leq t \leq 6$$

# 4 MEAN

The *mean* or *expected value* of a variable $X$ is the number, denoted by $E(X)$, obtained by one of the following:

(a) Multiplying the value of $X$ for each outcome by the probability of that outcome, and adding these products

(b) Multiplying each value of the variable by its probability of occurring, and adding these products

(a) $E(X) = \sum_{\text{outcomes } e} X(e)P(e)$

(b) $E(X) = \sum_{\text{values } v \text{ of } X} vP(X = v)$

# MEAN

The two formulas (a) and (b) are equivalent, since in the right-hand side of formula (a), the sum of all the terms with $X(e) = v$ is simply $vP(X = v)$.

**EXAMPLE 4.1** Two balls are selected at random without replacement from an urn containing three black balls and two white ones. Find $E(X)$ and $E(X^2)$, where $X$ is the number of black balls selected. Figure 4.1 provides a diagram.

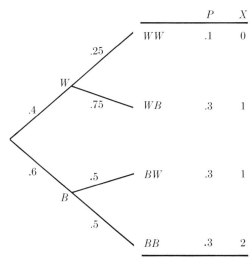

**FIGURE 4.1**

From formula (a):

$E(X) = 2(.3) + 1(.3) + 1(.3) + 0(.1) = 1.2$
$E(X^2) = 4(.3) + 1(.3) + 1(.3) + 0(.1) = 1.8$

To use formula (b), we find the distributions of $X$ and $X^2$:

$X$

| $v$ | $p$ |
|---|---|
| 0 | .1 |
| 1 | .6 |
| 2 | .3 |

$E(X) = 0(.1) + 1(.6) + 2(.3) = 1.2$

## BASIC STATISTICS

| $X^2$ | |
|---|---|
| $v$ | $p$ |
| 0 | .1 |
| 1 | .6 |
| 4 | .3 |

$E(X^2) = 0(.1) + 1(.6) + 4(.3) = 1.8$

**EXAMPLE 4.2** An urn contains two pennies, two nickels, and one dime. Two coins are selected at random, without replacement, and given to you. What is your expected income from the first draw? From the second draw? What is your total expected income? See Figure 4.2, where a diagram is given.

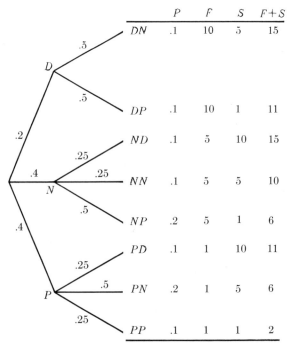

**FIGURE 4.2**

# MEAN

From formula (b):

$$E(F) = 1(.4) + 5(.4) + 10(.2) = 4.4$$
$$E(S) = 1(.4) + 5(.4) + 10(.2) = 4.4$$
$$E(F + S) = 2(.1) + 6(.4) + 10(.1) + 11(.2)$$
$$+ 15(.2) = 8.8$$

**DISCUSSION PROBLEM 4.1**

Suppose the drawing in Example 4.2 is made with replacement, and you are paid the sum of the values of the coins drawn. Is the distribution of your income the same as before? Is your expected income the same as before?

**Properties of the Mean**

1. The mean of a sum of variables is the sum of the means:

   $$E(X + Y + \cdots) = E(X) + E(Y) + \cdots$$

2. The mean of a constant times a variable is that constant times the mean of the variable:

   $$E(cX) = cE(X)$$

3. The mean of a constant is that constant:

   $$E(c) = c$$

The above properties follow at once from formula (a) for the mean.

**EXAMPLE 4.3**

A fair coin will be tossed repeatedly. Before each toss you can make an even-money bet of any amount on heads. Here are three systems, using only the first three tosses:

1. Bet $1 on the first toss. If you win, quit. If you lose, bet $2 on the second toss. If you win, quit. If you lose again, bet $4 on the third toss.
2. Bet $1 on the first toss. If you win, quit. If you lose, bet $1 on each of the second and third tosses.
3. Bet $1 on each toss.

Find the mean of your net income under each system.

| Outcome | Probability | $W_1$ | $W_2$ | $W_3$ |
|---------|-------------|-------|-------|-------|
| HHH | $\frac{1}{8}$ | 1 | 1 | 3 |
| HHT | $\frac{1}{8}$ | 1 | 1 | 1 |
| HTH | $\frac{1}{8}$ | 1 | 1 | 1 |
| HTT | $\frac{1}{8}$ | 1 | 1 | $-1$ |
| THH | $\frac{1}{8}$ | 1 | 1 | 1 |
| THT | $\frac{1}{8}$ | 1 | $-1$ | $-1$ |
| TTH | $\frac{1}{8}$ | 1 | $-1$ | $-1$ |
| TTT | $\frac{1}{8}$ | $-7$ | $-3$ | $-3$ |

From formula (a):

$$E(W_1) = \frac{1+1+1+1+1+1+1-7}{8} = \frac{0}{8} = 0$$

$$E(W_2) = 0$$

$$E(W_3) = 0$$

Your mean net income is 0 under each system.

**DISCUSSION PROBLEM 4.2** Invent another system to fit the circumstances described in Example 4.3, and find its mean net income.

**DISCUSSION PROBLEM 4.3** In system 3 of Example 4.3, what is the distribution of your net income from the first toss? From the second? From the third? What is your expected net income from each toss? What is your expected total net income? What would be your expected net income from system 3 if you used it for 10 tosses?

**EXAMPLE 4.4** Here is the density of a variable $X$:

$$p(t) = \begin{cases} t & 0 \leq t \leq 6 \\ 0 & \text{otherwise} \end{cases}$$

# MEAN

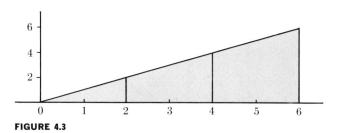
**FIGURE 4.3**

Find $E(X)$ and $E(X^2)$, approximately. See Figure 4.3. We use $X^*$ with values 1, 3, and 5. Its distribution is as follows:

| $v$ | $p$ |
|---|---|
| 1 | $\frac{1}{9}$ |
| 3 | $\frac{3}{9}$ |
| 5 | $\frac{5}{9}$ |

$E(X^*) = 1(\frac{1}{9}) + 3(\frac{3}{9}) + 5(\frac{5}{9}) = 3.89$

$E(X^{*2}) = 1(\frac{1}{9}) + 9(\frac{3}{9}) + 25(\frac{5}{9}) = \frac{153}{9} = 17$

Thus, approximately, $E(X) = 3.89$ and $E(X^2) = 17$. The exact values are $E(X) = 4$ and $E(X^2) = 18$.

An approximating $X^*$ with more values would give us a better approximation, with more work. For instance, an $X^*$ with six values .5, 1.5, . . . , 5.5, gives $E(X^*) = 3.97$ and $E(X^{*2}) = 17.86$.

## PROBLEMS Set 4

1. An urn contains three balls numbered 0, two balls numbered 2, and one ball numbered 4. Two balls are drawn successively at random from the urn, after which you are paid in dollars the average of the two numbers on the balls drawn. Find the distribution and mean of your income, and draw the histogram for your income (a) with replacement, or (b) without replacement.

Suppose after you see the first ball you can decide, depending on what it is, whether to replace it before the second draw. What is your best plan? What is your expected income if you use this plan?

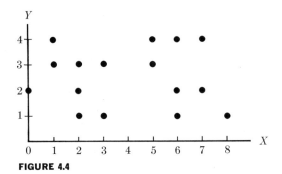

FIGURE 4.4

2. Another property of the mean is as follows: The mean of a variable is the point where its histogram or density balances. Use this fact to find $E(X)$ and $E(Y)$ in the scatter diagram of Figure 4.4. What is $E(X + Y)$? (Check by calculation.)

3. A variable $X$ has the following distribution:

| $v$ | $p$ |
|---|---|
| 0 | .4 |
| 1 | .3 |
| 2 | .2 |
| 3 | .1 |

Find the mean of the variable $(X - t)^2$ for $t = 0, 1, 2, 3$ and graph the mean of $(X - t)^2$ as a function of $t$ for $0 \leq t \leq 3$. What value of $t$ makes $E[(X - t)^2]$ smallest? What is $E(X)$?

4. One of the integers 1, 2, . . . , 1000 is selected at random. What is the mean value of the selected integer $X$ (where would the histogram balance)?

Since $E(X) = (1 + 2 + \cdots + 1000)/1000$, what is the sum of the first 1000 integers?

5. The distribution of a variable $X$ is as follows:

| $v$ | $p$ |
|---|---|
| 0 | $1 - t$ |
| 1 | $t$ |

Find (a) $E(X)$, (b) $E(X^2)$, (c) $[E(X)]^2$, (d) $E(X^2) - [E(X)]^2$, and graph each as a function of $t$, for $0 \leq t \leq 1$.

If a variable $X$ has only 0 and 1 as values, what are the possible values of $E(X)$? Of $E(X^2) - [E(X)]^2$?

6. If a variable $X$ has only 0, 1, and 2 as values, can you assign probabilities to these values so that $E(X) = 1$? So that $E(X^2) = \frac{5}{3}$? So that $E(X) = 1$ and $E(X^2) = \frac{5}{3}$? If $E(X) = 1$, what is the largest possible value of $E(X^2)$? The smallest possible value?

7. A fair coin is tossed until a tail occurs or until three tosses have been made, so that there are four possible outcomes: $T$, $HT$, $HHT$, and $HHH$. What is the expected number of tosses? What is the expected number of tosses if the limit 3 on the number of tosses is replaced by 2? By 4? By 5? By 1000 (guess)?

8. An urn contains one black ball and one white ball. Balls are drawn successively at random, with replacement, until the same color has occurred twice, so that the possible outcomes are $BB$, $BWB$, $BWW$, $WBB$, $WBW$, $WW$. Find the expected number of draws.

9. For each of the following densities, find $E(X)$ and $E(X^2)$, approximately.

(a) $f(t) = \begin{cases} 1 & 0 \leq t \leq 10 \\ 0 & \text{otherwise} \end{cases}$

(b) $f(t) = \begin{cases} 12 - t & 0 \leq t \leq 12 \\ 0 & \text{otherwise} \end{cases}$

10. For a certain variable $X$, Figure 4.5 shows the graph of $P(X \leq t)$. Find $E(X)$ and $E(X^2)$, approximately.

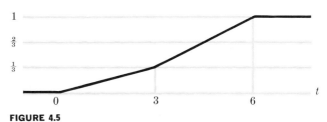

FIGURE 4.5

11. A coin with probability $p$ of heads will be tossed twice. On each toss you may make an even-money bet of any amount on heads. You use the following system: Bet \$2 on the first toss. If you win bet \$1 on the second toss, but if you lose bet \$4 on the second toss.

(a) Find the distribution and mean of your net winnings $W$ for $p = .4, .5,$ and $.6$.
(b) Denote the above system by (2; 1,4). Find the mean of $W$ for $p = .4$ and $.6$ for each of these systems: (1; 2,3), (1; 0,2), (1; 0,1).
(c) Choose any system $(a; b,c)$ and find $E(W)$ for $p = .4$ and $.6$. Formulate a general conclusion about the sign ($\pm$) of $E(W)$.

12. In a certain population, 60% of the voters are Democrat. You select five voters at random with replacement from the population, and give each Democrat a dollar. What is your expected payment to the first voter selected? To the fourth? What is your expected total payment? Answer these questions for selection without replacement.

13. You have five customers, two at $A$, two at $B$, and one at $C$. You may locate yourself anywhere on the line segment $AC$ of Figure 4.6. Every day one of the cus-

FIGURE 4.6

tomers is selected at random, and you must visit the selected customer. Where should you locate yourself to minimize the expected distance traveled? Suppose the cost of travel is the square of the distance traveled, so that, for instance, if you are at 3, a visit to $C$ costs you $(5 + 5)^2 = 100$ dollars. Where should you locate yourself to minimize your expected cost?

14. A point is selected at random inside a circle of radius 1. What is the probability that the distance $X$ of the selected point from the center is at least .5? Find $E(X)$, approximately.

15. An urn contains three balls labeled 0, 1, and 2. Two balls are drawn successively at random. Find $E(XY)$, where $X$ and $Y$ are the first and second numbers drawn (a) with replacement, or (b) without replacement.

16. In Problem 15, find the expected value of $Z = \max(X, Y)$, that is, of the larger of the two numbers drawn (if $X$ equals $Y$, then $Z$ equals their common value).

17. Every time a coin falls heads, your fortune is doubled; every time it falls tails, your fortune is halved. You start with $1. What is your expected fortune after two tosses (a) if the coin is fair, and (b) if the probability of heads is $\frac{1}{3}$?

# 5 VARIANCE

You have a job as a predictor. One of the three words in the sentence GO TO LUNCH will be selected at random. You must predict the number of letters in the selected word, and the square of your error will be deducted from your salary. For instance, if you predict 3 and the word LUNCH is selected, you lose $(5 - 3)^2 = 4$ dollars, while if GO or TO is selected, you lose $(2 - 3)^2 = 1$ dollar. What number should you predict to make your expected loss as small as possible, and how much is this smallest expected loss?

If you choose the number $t$ as your prediction, here is the calculation of your expected loss:

| Word | $L$ (length) | $L - t$ (error) | $(L - t)^2$ (squared error, or loss) |
|---|---|---|---|
| GO | 2 | $2 - t$ | $(2 - t)^2$ |
| TO | 2 | $2 - t$ | $(2 - t)^2$ |
| LUNCH | 5 | $5 - t$ | $(5 - t)^2$ |

$$E(\text{loss}) = E(L - t)^2 = \tfrac{1}{3}(2 - t)^2 + \tfrac{1}{3}(2 - t)^2 + \tfrac{1}{3}(5 - t)^2$$
$$= \tfrac{1}{3}(33 - 18t + 3t^2)$$
$$= t^2 - 6t + 11$$
$$= (t - 3)^2 + 2$$

Your expected loss (expected squared error, or mean squared error, mse) will be at least 2 for every $t$, and will be exactly 2 if $t = 3$. You should predict 3 for $L$, and your mse will then be 2. Notice that

$$E(L) = \frac{2 + 2 + 5}{3} = 3$$

the same as your best prediction.

> For any variable $X$, the constant $t$ that makes the mean of $(X - t)^2$ smallest is $t = E(X)$. For this choice of $t$, the mean of $(X - t)^2$, which is called the *variance* of $X$ and is denoted by $\sigma^2(X)$, is given by the formula
> 
> $$\sigma^2(X) = E(X^2) - [E(X)]^2$$

To see that $E(X)$ is the best choice for $t$ and that the formula given for $\sigma^2(X)$ is correct,

$$\begin{aligned}E(X-t)^2 &= E(X^2 - 2tX + t^2)\\ &= E(X^2) - 2tE(X) + t^2\\ &= [t - E(X)]^2 + E(X^2) - [E(X)]^2\end{aligned}$$

The nonnegative square root of the variance of $X$ is called the *standard deviation* of $X$ and is denoted by $\sigma(X)$.

**EXAMPLE 5.1** Two people are selected at random with replacement from a population that is 60% male. Find the standard deviation of $X$, the number of males in the selected sample.

| Sample | Probability | $X$ | $X^2$ |
|--------|-------------|-----|-------|
| $MM$ | $(.6)(.6) = .36$ | 2 | 4 |
| $MF$ | $(.6)(.4) = .24$ | 1 | 1 |
| $FM$ | $(.4)(.6) = .24$ | 1 | 1 |
| $FF$ | $(.4)(.4) = .16$ | 0 | 0 |

$$E(X) = 2(.36) + 1(.48) + 0(.16) = 1.2$$
$$E(X^2) = 4(.36) + 1(.48) + 0(.16) = 1.92$$
$$\sigma^2(X) = 1.92 - (1.2)^2 = .48$$
$$\sigma(X) = \sqrt{.48} = .69$$

We can find from the table of squares at the end of the book that $(.69)^2 = .4761$ and $(.70)^2 = .4900$.

From our interpretation of variance, if you had to predict the number of males in the sample, losing the square of your error, your best prediction would be 1.2, and your expected loss would be .48.

**Properties of Variance and Standard Deviation**

1. Adding a constant to a variable does not change the variance or the standard deviation:

$$\sigma^2(X + c) = \sigma^2(X)$$
$$\sigma(X + c) = \sigma(X)$$

2. Multiplying a variable by a constant multiplies the variance by the square of the constant, and the standard deviation by the absolute value of that constant:

$$\sigma^2(cX) = c^2\sigma^2(X)$$
$$\sigma(cX) = |c|\sigma(X)$$

The two properties given above may be expressed in a single statement:

For any constants $a$ and $b$,

$$\sigma^2(aX + b) = a^2\sigma^2(X)$$
$$\sigma(aX + b) = |a|\sigma(X)$$

One use of these properties is to simplify calculation.

**EXAMPLE 5.2** Following is the distribution of weight ($W$) in a population. Find $E(W)$ and $\sigma(W)$.

| $v$ | $p$ |
|---|---|
| 140 | .5 |
| 150 | .3 |
| 160 | .2 |

We look instead at $X = (W - 150)/10$, whose distribution is as follows:

| $v$ | $p$ |
|---|---|
| −1 | .5 |
| 0 | .3 |
| 1 | .2 |

$$E(X) = -1(.5) + 0(.3) + 1(.2) = -.3$$
$$E(X^2) = 1(.5) + 0(.3) + 1(.2) = .7$$
$$\sigma^2(X) = (.7) - (-.3)^2 = .61$$
$$\sigma(X) = \sqrt{.61} = .78$$

Since

$$W = 10X + 150$$
$$E(W) = 10(-.3) + 150 = 147$$
$$\sigma^2(W) = 10^2(.61) = 61$$
$$\sigma(W) = 10(.78) = 7.8$$

The standard deviation of a variable is the most common measure of dispersion (spread or variability) of the variable. For many variables, the probability that the variable is within one standard deviation of its mean is between 60% and 70%. Figure 5.1 shows 16 densities on $0 \le t \le 1$. For each, the region between $m - \sigma$ and $m + \sigma$ is shaded, and the probability $p$ in this region (that is, the ratio of the shaded area to the total area under the density) is shown.

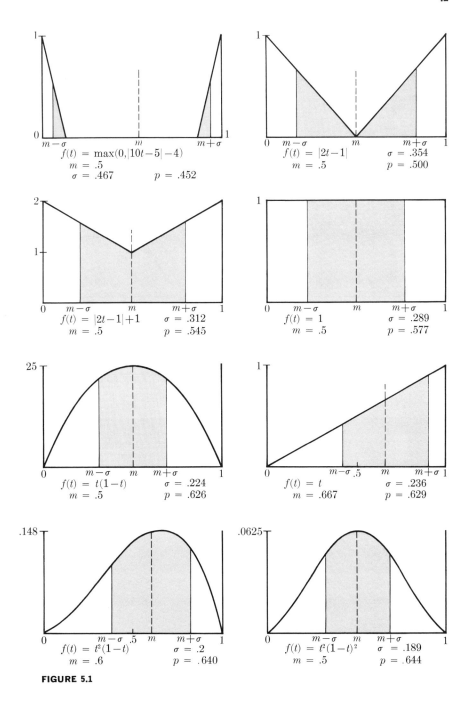

**FIGURE 5.1**

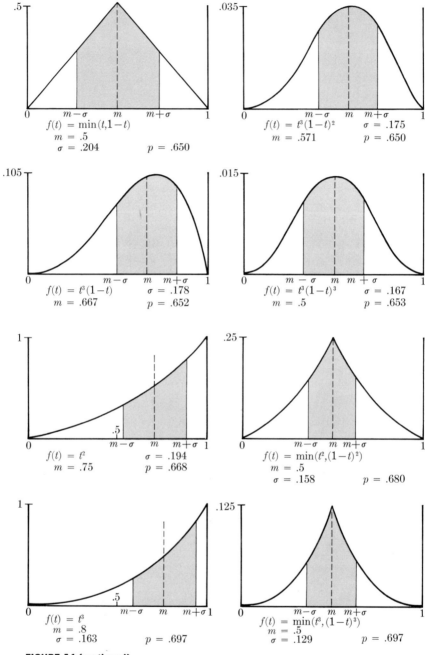

**FIGURE 5.1 (continued)**

# VARIANCE

**PROBLEMS**
**Set 5**

1. In each scatter diagram shown in Figure 5.2, estimate the mean and standard deviation of $X$ and $Y$. Then calculate the correct values and mark in each scatter diagram the point $(E(X), E(Y))$. For each scatter diagram find $\sigma(X + Y)$.

2. One of the digits 0, . . . , 9 is selected at random. Find the mean and standard deviation of the selected digit.

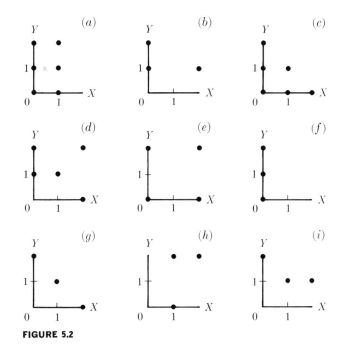

FIGURE 5.2

3. An urn contains two white balls and two black balls. Three balls are drawn at random; find the mean and standard deviation of the number of white balls drawn (a) with replacement, or (b) without replacement.

4. Room 1 contains three women whose heights are 62, 65, and 65. Room 2 contains two men whose heights are 70 and 74. One of the two rooms is selected at random,

then a person is selected at random from the selected room. You must predict the height of the person selected, losing the square of your error. What is your best prediction, and what is your mean squared error? Suppose that, before making your prediction, you are told which room was chosen. What is your best prediction given room 1, and what is your mse? Given room 2? Suppose someone offers, for a price, to tell you which room was chosen. How much would you be willing to pay?

5.  Balls are drawn successively at random with replacement from an urn in which 60% of the balls are black.

(a) Let $S_n$ denote the number of black balls among the first $n$ drawn. Find $E(S_n)$ and $\sigma^2(S_n)$ for $n = 1, 2, 3,$ and 4. Try to guess a general formula for $E(S_n)$ and for $\sigma^2(S_n)$.

(b) Let $Y_n$ denote the proportion of black balls among the first $n$ drawn, so that $Y_n = S_n/n$. Find $E(Y_n)$ and $\sigma^2(Y_n)$ for $n = 1, 2, 3,$ and 4. Try to guess a general formula for $E(Y_n)$ and for $\sigma^2(Y_n)$.

6.  Repeat Problem 5, but with 60% replaced by 30%.

7.  With 60% replaced by $p$ in Problem 5, try to guess general formulas for $E(S_n)$, $\sigma^2(S_n)$, $E(Y_n)$, and $\sigma^2(Y_n)$.

8.  Below are some incompletely specified distributions. In each case, try to find a value of $t$ that will make the standard deviation as large as possible, and one that will make it as small as possible. Draw the histograms for the selected values of $t$.

(a)

| $v$ | $p$ |
|---|---|
| 0 | $t$ |
| 1 | $1-t$ |

(b)

| $v$ | $p$ |
|---|---|
| 1 | $t$ |
| 6 | $1-t$ |

(c)

| $v$ | $p$ |
|---|---|
| 0 | $t$ |
| 1 | .4 |
| 2 | $.6 - t$ |

(d)

| $v$ | $p$ |
|---|---|
| 0 | $t$ |
| 1 | $.6 - t$ |
| 2 | .4 |

(e)

| $v$ | $p$ |
|---|---|
| 0 | $\frac{1}{3}$ |
| 2 | $\frac{1}{3}$ |
| $t$ | $\frac{1}{3}$ |

($t$ between 0 and 2)

(f)

| $v$ | $p$ |
|---|---|
| 0 | $\frac{1}{3}$ |
| 2 | $\frac{1}{3}$ |
| $t$ | $\frac{1}{3}$ |

($t$ unrestricted)

(g)

| $v$ | $p$ |
|---|---|
| $t$ | .5 |
| $t+1$ | .3 |
| $t+2$ | .2 |

(h)

| $v$ | $p$ |
|---|---|
| $t$ | .5 |
| $2t$ | .3 |
| $3t$ | .2 |

($t$ between 1 and 2)

9. Graph the densities represented by each of the following functions on $0 \le t \le 1$, estimate m, $\sigma$, and $p = P(m - \sigma \le X \le M + \sigma)$ from your graph, and make approximate calculations to check your estimates.

(a) $f(t) = \dfrac{t}{t+1}$  (b) $f(t) = (2t-1)^2$

10. A point is selected at random inside a circle of radius 6 feet. Find approximately the standard deviation of $X$, the distance of the selected point from the center, by using an approximating $X^*$ with values 1, 3, and 5.

11. A fair coin is tossed until tails occurs or until three tosses have been made. Find the standard deviation of the number $N$ of tosses.

# 6 WORTH OF A PREDICTOR

When you had to predict the length of $L$ of a word selected at random from the sentence GO TO LUNCH, losing the square of your error, your best prediction of $L$ was $E(L) = 3$, and your expected loss or mean squared error using this best prediction was

$$\sigma^2(L) = E(L-3)^2 = 2$$

Suppose before predicting $L$ you are told the number $X$ of $G$'s in the selected word. Your best prediction for $L$, using this information, is clearly 2 when $X = 1$, since you know that GO was selected. When $X = 0$, you know that TO or LUNCH was selected, and they are equally probable, so your best prediction is

$$E(L|X = 0) = \tfrac{1}{2}(2) + \tfrac{1}{2}(5) = 3.5$$

Your prediction $Y$ is now a variable whose value depends on $X$, or on which word was selected. We calculate your mse:

| Word | $L$ | $X$ | $Y$ | $L - Y$ (error) | $(L - Y)^2$ (squared error) |
|---|---|---|---|---|---|
| GO | 2 | 1 | 2 | 0 | 0 |
| TO | 2 | 0 | 3.5 | $-1.5$ | 2.25 |
| LUNCH | 5 | 0 | 3.5 | 1.5 | 2.25 |

$$\text{mse for } Y \text{ as predictor of } L = E(L - Y)^2$$
$$= \frac{0 + 2.25 + 2.25}{3} = 1.5$$

Using $Y$ to predict $L$ gives an expected saving, or an expected reduction in squared error, of .50 from the original mse of 2. Your expected saving is 25% of the original expected cost. We say that $Y$ is *worth* .25 as a predictor of $L$.

For any two variables $Y$ and $L$:

(a) The mse of $Y$ as a predictor of $L$ is
$$E(L - Y)^2$$

(b) The *worth* of $Y$ as a predictor of $L$ is
$$W(Y,L) = \frac{\sigma^2(L) - E(L - Y)^2}{\sigma^2(L)}$$

The variable that has the highest worth for predicting $L$ is $L$ itself, which has worth 1:

$$W(L,L) = \frac{\sigma^2(L) - E(L-L)^2}{\sigma^2(L)} = 1$$

The mean of $L$ has worth 0 as a predictor of $L$:

$$W(E(L), L) = \frac{\sigma^2(L) - E(L - E(L))^2}{\sigma^2(L)}$$

$$= \frac{\sigma^2(L) - \sigma^2(L)}{\sigma^2(L)} = 0$$

Some predictors have negative worth; for instance, any constant other than $E(L)$ has negative worth as a predictor of $L$.

**EXAMPLE 6.1** Figure 6.1 is a scatter diagram for two variables $X$ and $Y$. Find the mse and worth of $Z = .1X + 1$ as a predictor of $Y$.

The line is the graph of $Z$ as a function of $X$. The calculations are as follows:

|   | $X$ | $Y$ | $Z$ | $(Y-Z)^2$ | $Y^2$ |
|---|---|---|---|---|---|
|   | 0 | 0 | 1   | 1    | 0   |
|   | 0 | 1 | 1   | 0    | 1   |
|   | 1 | 2 | 1.1 | .81  | 4   |
|   | 3 | 1 | 1.3 | .09  | 1   |
| Σ | 4 | 4 | 4.4 | 1.90 | 6   |
| E | 1 | 1 | 1.1 | .475 | 1.5 |

$$\sigma^2(Y) = 1.5 - 1^2 = .5$$

$$W(Z,Y) = \frac{.5 - .475}{.5} = .05$$

Thus $Z$ has mse .475 and worth .05 as a predictor of $Y$.

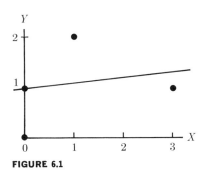

**FIGURE 6.1**

**DISCUSSION PROBLEM 6.1**

What constant $b$ makes $E(Y - Z - b)^2$ smallest? For this $b$ does $Z + b$ have smaller mse than $Z$ as a predictor of $Y$? Calculate $W(Z + b, Y)$ for this $b$.

**DISCUSSION PROBLEM 6.2**

What is the best function of $X$ for predicting $Y$? What is its worth? What is the best function of $Y$ for predicting $X$? What is its worth?

**PROBLEMS Set 6**

1. Figure 6.2 shows the scatter diagram for two variables $X$ and $Y$.

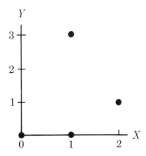

**FIGURE 6.2**

(a) Find the mse of each of the following constants for predicting Y, and the worth of each: 0, .5, 1, 1.5, 2.

(b) Find the mse and worth of each of the following functions of $X$ for predicting $Y$: $X$, $2X$, $2X - 1$, $.8X + .2$.

(c) Find the best function of $X$ for predicting $Y$, and its worth.

(d) Each of the functions in (b) is a linear function of $X$, that is, has the form $aX + b$, where $a$ and $b$ are constants. Can you find constants $a$ and $b$ for which

$aX + b$ is a better predictor of $Y$ (has higher worth) than any of the functions in (b)?

(e) Find the best function of $Y$ for predicting $X$, and its worth. Is it a linear function of $Y$?

2. A fair coin is tossed three times. Denote by $X$ the number of heads in the first two tosses and by $Y$ the number of heads in the three tosses. Draw the scatter diagram for $X$ and $Y$. (Remember to label each point with a number proportional to its probability.)

(a) Find the best function of $X$ for predicting $Y$. What is its worth? Is it a linear function of $X$?

(b) Find the best function of $Y$ for predicting $X$. What is its worth? Is it a linear function of $Y$?

3. Repeat Problem 2, except that the probability of heads in each toss is now .6.

4. In each scatter diagram of Figure 6.3, find the worth of the best function of $X$ for predicting $Y$ and of the best function of $Y$ for predicting $X$.

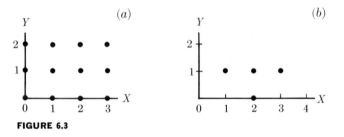

FIGURE 6.3

5. Draw a scatter diagram, with a small number of points, for two variables $X$ and $Y$, so that $E(X) = 1$ and $E(Y) = 3$. Suppose you are going to use the variable $2X + c$, where $c$ is some constant, to predict $Y$. What would be the best choice for $c$? Calculate your mse for $c = 0, 1, 2,$ and $3$.

# 7 CORRELATION

You have to predict the length $Z$ of a word selected at random from the sentence I SEE THE MOUSE, losing the square of your error. Your best constant predictor is $E(Z) = 3$, and your mse is then 2, since you lose 0 and 4 with probability $\frac{1}{2}$ each, $2 = \sigma^2(Z)$.

If, before predicting $Z$, you are told the number $X$ of $E$'s in the selected word, we see from the scatter diagram for $X$ and $Z$ in Figure 7.1 that the best predictions when $X = 0$, 1, and 2 are 1, 4, and 3, respectively; the circled points are the graph of this best predictor, as a function of $X$. Using this predictor you lose 0 and 1 with probability $\frac{1}{2}$ each, so your mse is .5, and the predictor is worth $(2 - .5)/2 = .75$.

The three circled points do not lie on a line, and thus the best function of $X$ for predicting $Z$ is not a linear function of $X$, or is not of the form $aX + b$, where $a$ and $b$ are constants. What is the best linear function of $X$ for predicting $Z$, and what is its worth?

---

For any two variables $X$ and $Y$, the *covariance* of $X$ and $Y$ is the number

$$\text{cov}(X,Y) = E(XY) - E(X)E(Y)$$

The best linear function of $X$ for predicting $Y$ is

$$U = aX + b$$

where

$$a = \frac{\text{cov}(X,Y)}{\text{cov}(X,X)} \quad \text{and} \quad b = E(Y) - aE(X)$$

The worth of this best linear function of $X$ for predicting $Y$ is denoted by $\rho^2(X,Y)$. It is called the *squared correlation coefficient* between $X$ and $Y$, and is given by the formula

$$\rho^2(X,Y) = \frac{\text{cov}^2(X,Y)}{\text{cov}(X,X)\text{cov}(Y,Y)}$$

The square root of $\rho^2$, with the sign ($\pm$) of $a$, is called the *correlation coefficient* between $X$ and $Y$, and is denoted by $\rho(X,Y)$.

---

Note that $\text{cov}(X,Y) = \text{cov}(Y,X)$, so that $\rho(X,Y) = \rho(Y,X)$, and that $\text{cov}(X,X) = \sigma^2(X)$.

# CORRELATION

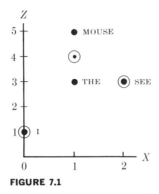

**FIGURE 7.1**

The calculations for finding the best linear function of $X$ for predicting $Z$, and its worth, are as follows:

| Word | $X$ | $Z$ | $X^2$ | $XZ$ | $Z^2$ | $U = X + 2$ | $Z - U$ (error) | $Z - U^2$ (squared error) |
|---|---|---|---|---|---|---|---|---|
| I     | 0 | 1 | 0 | 0  | 1  | 2  | −1 | 1   |
| SEE   | 2 | 3 | 4 | 6  | 9  | 4  | −1 | 1   |
| THE   | 1 | 3 | 1 | 3  | 9  | 3  | 0  | 0   |
| MOUSE | 1 | 5 | 1 | 5  | 25 | 3  | 2  | 4   |
| Σ     | 4 | 12| 6 | 14 | 44 | 12 | 0  | 6   |
| E     | 1 | 3 | 1.5 | 3.5 | 11 | 3 | 0 | 1.5 |

$$\text{cov}(X,X) = 1.5 - (1)^2 = .5$$
$$\text{cov}(X,Z) = 3.5 - 1(3) = .5$$
$$\text{cov}(Z,Z) = 11 - (3)^2 = 2$$
$$a = \frac{\text{cov}(X,Z)}{\text{cov}(X,X)} = \frac{.5}{.5} = 1$$
$$b = E(Z) - aE(X) = 3 - 1(1) = 2$$
$$\rho^2(X,Z) = \frac{\text{cov}^2(X,Z)}{\text{cov}(X,X)\text{cov}(Z,Z)} = \frac{(.5)^2}{(.5)(2)} = .25$$

The best linear function of $X$ for predicting $Z$ is $U = X + 2$, and its worth is $\rho^2(X,Z) = .25$.

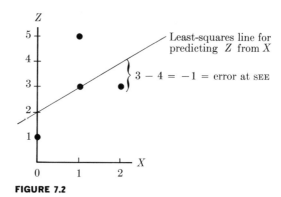
**FIGURE 7.2**

Figure 7.2 shows the graph of $U$ as a function of $X$. Since $U$ is linear, its graph is a line, called the *least-squares line for predicting Z from X*.

Remarks

1. The last three columns of the table are calculated after $a$ and $b$ are calculated.

2. The mean of $U$ is the same as the mean of the variable $Z$ it is predicting, so $E(Z - U) = 0$. This will always be true for the best linear predictor, so the calculation checks our work.

3. We found the mse of $U$ as a predictor of $Z$ to be 1.5, so $W(U,Z) = (2 - 1.5)/2 = .25$, agreeing with $\rho^2(X,Z)$ and checking our work again.

4. We could have had an mse of only .5, worth .75, instead of mse 1.5, worth .25, if we had used the best function of $X$ for predicting $Z$ instead of the best *linear* function.

5. The least-squares line can be drawn by calculating any two prediction points, that is, any two pairs of values $(X,U)$ plotting them, and joining them with a line; for example, when $X = 0$, $U = 2$ and when $X = 2$, $U = 4$, so the least-squares line joins (0,2) and (2,4).

Another point that is always on the least-squares line is $(E(X), E(Z))$. Here $(E(X), E(Z)) = (1,3)$.

6. The error associated with each point on the scatter diagram is the vertical distance from the least-squares line to that point.

7. The number $a$ measures how much our prediction of $Z$ increases when $X$ increases by 1. It is called the *slope* of the least-squares line. When $a = 0$, the best line is horizontal, that is, the best linear predictor is constant, and thus worth 0: $\rho^2(X,Z) = 0$.

**DISCUSSION PROBLEM 7.1** In the example we have been studying, what is the largest squared error we could have in using $U$ to predict $Z$? Can you find another linear function of $X$ whose largest squared error in predicting $Z$ is less than this?

**DISCUSSION PROBLEM 7.2** Figure 7.3 shows a scatter diagram. See if you can answer the following questions without writing anything; then do the calculations to check your answers:

(a) What is the best horizontal line for predicting $Y$?
(b) What is its mean squared error as a predictor of $Y$, that is, what is the variance of $Y$?
(c) What is the best function of $X$ for predicting $Y$?
(d) What is the least-squares line for predicting $Y$ from $X$?
(e) What is its mse as a predictor of $Y$?
(f) What is the worth of the least-squares line for predicting $Y$, that is, what is $\rho^2(X,Y)$?

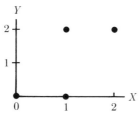

**FIGURE 7.3**

## PROBLEMS
## Set 7

1. A word is selected at random from the sentence I SEE THE MOUSE. $X$ is the number of $E$'s in the selected word, $Y$ is the number of $I$'s in the selected word, and $Z$ is the length of the selected word.

   Draw the scatter diagram for each of the following pairs, find $\rho^2$ and the best linear function of the first for predicting the second, and draw the least-squares line:

   (a) $(Z,X)$
   (b) $(Y,Z)$
   (c) $(X, Z - X)$
   (d) $(X + 4Y, Z)$
   (e) $(X^2, Z)$

2. Take any scatter diagram for two variables $X$ and $Y$, find the best linear function $U$ of $X$ for predicting $Y$, and check the following facts:

   (a) $\sigma^2(U) + \sigma^2(Y - U) = \sigma^2(Y)$
   (b) $\rho^2(U, Y - U) = 0$
   (c) The best linear function of $X$ for predicting $2Y + 3$ is $2U + 3$.
   (d) $\text{cov}(X - 1, Y + 2) = \text{cov}(X,Y)$ (Adding constants to variables does not change their covariance.)
   (e) $\text{cov}(2X, -3Y) = -6 \text{ cov}(X,Y)$ (Multiplying variables by constants multiplies their covariance by the product of these constants.)
   (f) $\sigma^2(X + Y) = \sigma^2(X) + \sigma^2(Y) + 2 \text{ cov}(X,Y)$
   (g) $\rho^2(2X - 1, Y) = \rho^2(X,Y)$ (Formulate this property.)

3. For any three variables $X$, $Y$ and $Z$, verify $\text{cov}(X, Y + Z) = \text{cov}(X,Y) + \text{cov}(X,Z)$.

4. In each scatter diagram of Figure 7.4, add a fourth point (a) to make $\rho^2(X,Y) = 0$, or (b) to make $\rho^2(X,Y)$ as large as possible.

# CORRELATION

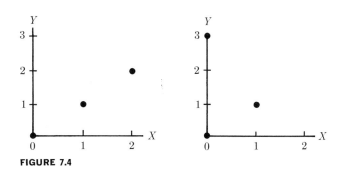

**FIGURE 7.4**

5. A scatter diagram for $X$ and $Y$ has three points: $(-1,0)$, $(1,0)$, and a third point $Q = (h,k)$. Find the best linear function $U = aX + b$ of $X$ for predicting $Y$ and $\rho^2(X,Y)$, for each of the nine points $Q$ with $h,k = 1$, 3, and 10, and show the results in tables:

| $a$ | | | | $b$ | | | | $\rho^2$ | | | |
|---|---|---|---|---|---|---|---|---|---|---|---|
| h \ k | 1 | 3 | 10 | h \ k | 1 | 3 | 10 | h \ k | 1 | 3 | 10 |
| 1 | | | | 1 | | | | 1 | | | |
| 3 | | | | 3 | | | | 3 | | | |
| 10 | | | | 10 | | | | 10 | | | |

For fixed $h$, how does $a$ depend on $k$ (increases, decreases, remains constant)? For fixed $h$, how do $b$ and $\rho^2$ depend on $k$? For fixed $k$, how do $a$, $b$, and $\rho^2$ depend on $h$?

Find a point $Q$ with $.4 \leq \rho^2 \leq .6$ and $0 < a < .5$. Draw the scatter diagram and the least-squares line for this $Q$.

6. The following table shows the fraction of the total vote received by a candidate for governor and a candidate for senator in each of four precincts:

| Precinct | G | S |
|---|---|---|
| 1 | .5 | .4 |
| 2 | .8 | .6 |
| 3 | .3 | .3 |
| 4 | .4 | .3 |

Find the best linear function of $G$ for predicting $S$, and $\rho^2(G,S)$. (It will simplify the calculations if you use $X = G - .5$ and $Y = S - .3$).

7. Urn 1 contains 40% black balls and urn 2 contains 80% black balls. One of the two urns is selected at random, then a ball is drawn at random from the selected urn. Find the best linear function of $X$, the number of black balls drawn, for predicting $Y$, the proportion of black balls in the selected urn, and find $\rho^2(X,Y)$.

8. In a certain population, 50% of the people own no car, and 50% own one car. Any person, when asked how many cars he owns, tells the truth with probability .8 and lies with probability .2. Find the correlation between $X$, the answer given, and $Y$, the number of cars owned.

# 8 MULTIPLE AND PARTIAL CORRELATION

You have to predict the length $Z$ of a word selected at random from the sentence I SEE THE MOUSE, losing the square of your error. Before predicting $Z$, you will be given the values of the two variables

$X$ = number of $E$'s in the selected word
$Y$ = number of $I$'s in the selected word

but must use a linear function

$V = cX + dY + e \quad c, d,$ and $e$ constant

of $X$ and $Y$ to predict $Z$. What is the best linear function $V$, and what is its worth?

The best linear function of two variables $X$ and $Y$ for predicting a third variable $Z$ is

$$V = cX + dY + e$$

where

$$c = \frac{\operatorname{cov}(X,Z)\operatorname{cov}(Y,Y) - \operatorname{cov}(Y,Z)\operatorname{cov}(X,Y)}{D}$$

$$d = \frac{\operatorname{cov}(Y,Z)\operatorname{cov}(X,X) - \operatorname{cov}(X,Z)\operatorname{cov}(Y,X)}{D}$$

$$e = E(Z) - cE(X) - dE(Y)$$

$$D = \operatorname{cov}(X,X)\operatorname{cov}(Y,Y) - \operatorname{cov}^2(X,Y)$$

The worth of $V$ for predicting $Z$, called the *squared multiple correlation coefficient between $Z$ and $(X,Y)$* and denoted by $\rho^2(Z,(X,Y))$ is

$$\rho^2(Z,(X,Y)) = \frac{c\operatorname{cov}(X,Z) + d\operatorname{cov}(Y,Z)}{\operatorname{cov}(Z,Z)}$$

The amount by which $\rho^2(Z,(X,Y))$ exceeds $\rho^2(Z,X)$, divided by the largest possible increase $1 - \rho^2(Z,X)$, is called the *squared partial correlation coefficient between $Y$ and $Z$ given $X$*, and is denoted by $\rho^2(Y,Z|X)$:

$$\rho^2(Y,Z|X) = \frac{\rho^2(Z,(X,Y)) - \rho^2(Z,X)}{1 - \rho^2(Z,X)}$$

The square root of $\rho^2(Y,Z|X)$, with the sign ($\pm$) of $d$, is called the *partial correlation coefficient between $Y$ and $Z$ given $X$*, and is denoted by $\rho(Y,Z|X)$.

The calculations for our example are as follows:

| Word | X | Y | Z | $X^2$ | $Y^2$ | $Z^2$ | XY | XZ | YZ | V | $(Z-V)$ | $(Z-V)^2$ |
|---|---|---|---|---|---|---|---|---|---|---|---|---|
| I     | 0 | 1 | 1 | 0 | 1 | 1  | 0 | 0  | 1 | 1  | 0  | 0 |
| SEE   | 2 | 0 | 3 | 4 | 0 | 9  | 0 | 6  | 0 | 3  | 0  | 0 |
| THE   | 1 | 0 | 3 | 1 | 0 | 9  | 0 | 3  | 0 | 4  | −1 | 1 |
| MOUSE | 1 | 0 | 5 | 1 | 0 | 25 | 0 | 5  | 0 | 4  | 1  | 1 |
| Σ     | 4 | 1 | 12 | 6 | 1 | 44 | 0 | 14 | 1 | 12 | 0  | 2 |
| E     | 1 | $\tfrac{1}{4}$ | 3 | $\tfrac{3}{2}$ | $\tfrac{1}{4}$ | 11 | 0 | $\tfrac{7}{2}$ | $\tfrac{1}{4}$ | 3 | 0 | $\tfrac{1}{2}$ |

**COVARIANCE TABLE**

|   | X | Y | Z |
|---|---|---|---|
| X | $\tfrac{1}{2}$ | $-\tfrac{1}{4}$ | $\tfrac{1}{2}$ |
| Y | $-\tfrac{1}{4}$ | $\tfrac{3}{16}$ | $-\tfrac{1}{2}$ |
| Z | $\tfrac{1}{2}$ | $-\tfrac{1}{2}$ | 2 |

$$D = \text{cov}(X,X)\text{cov}(Y,Y) - \text{cov}^2(X,Y)$$
$$= \tfrac{1}{2}(\tfrac{3}{16}) - (-\tfrac{1}{4})^2 = \tfrac{1}{32}$$

$$c = \frac{\text{cov}(X,Z)\text{cov}(Y,Y) - \text{cov}(Y,Z)\text{cov}(X,Y)}{D}$$
$$= \frac{\tfrac{1}{2}(\tfrac{3}{16}) - (-\tfrac{1}{2})(-\tfrac{1}{4})}{\tfrac{1}{32}} = -1$$

$$d = \frac{\text{cov}(Y,Z)\text{cov}(X,X) - \text{cov}(X,Z)\text{cov}(Y,X)}{D}$$
$$= \frac{-\tfrac{1}{2}(\tfrac{1}{2}) - \tfrac{1}{2}(-\tfrac{1}{4})}{\tfrac{1}{32}} = -4$$

$$e = E(Z) - cE(X) - dE(Y)$$
$$= 3 - (-1)(1) - (-4)(\tfrac{1}{4}) = 5$$

$$\rho^2(Z,(X,Y)) = \frac{-1(\tfrac{1}{2}) + (-4)(-\tfrac{1}{2})}{2} = \frac{3}{4}$$

$$\rho^2(Z,X) = \frac{(\tfrac{1}{2})^2}{\tfrac{1}{2}(2)} = \frac{1}{4}$$

# BASIC STATISTICS

$$\rho^2(Z,Y) = \frac{(-\frac{1}{2})^2}{\frac{3}{16}(2)} = \frac{2}{3}$$

$$\rho^2(Y,Z|X) = \frac{\frac{3}{4} - \frac{1}{4}}{1 - \frac{1}{4}} = \frac{2}{3}$$

$$\rho^2(X,Z|Y) = \frac{\frac{3}{4} - \frac{2}{3}}{1 - \frac{2}{3}} = \frac{1}{4}$$

The best linear function of $X$ and $Y$ for predicting $Z$ is $V = -X - 4Y + 5$ and its worth is $\rho^2(Z,(X,Y)) = \frac{3}{4}$.

**Remarks**

1. $E(Z - V) = E((Z - V)X) = E((Z - V)Y) = 0$
The error $(Z - V)$ has mean 0 and is uncorrelated with $X$ and $Y$. This will always be true.

2. We can calculate the worth of $V$ as a predictor of $Z$ directly:

$$W(V,Z) = \frac{2 - \frac{1}{2}}{2} = \frac{3}{4}$$

3. The coefficient $c$ of $X$ is negative. An increase in $X$ for fixed $Y$ decreases our prediction of $Z$, in contrast with the best linear function of $X$ alone for predicting $Z$, found earlier to be $X + 2$, where increasing $X$ increases our prediction of $Z$.

**PROBLEMS Set 8**

1. Following are the values of three variables $X$, $Y$, and $Z$ on a population with five elements:

| Element | X | Y | Z |
|---------|---|---|---|
| a | 0 | 1 | 0 |
| b | 0 | 1 | 1 |
| c | 1 | 2 | 2 |
| d | 2 | 2 | 2 |
| e | 2 | 4 | 4 |

One of the five elements is selected at random. Find the best linear function $U$ of $X$ for predicting $Z$, the best linear function $V$ of $X$ and $Y$ for predicting $Z$, the best linear function $W$ of $X$ for predicting $Y$, and $\rho^2(X,Z)$, $\rho^2(Y,Z)$, $\rho^2(Z,(X,Y))$, $\rho^2(Y, Z|X)$, and $\rho^2(Z, X|Y)$.

2. For any three variables $X$, $Y$, and $Z$, the partial correlation between $Y$ and $Z$ given $X$ is the correlation between the errors in the best linear functions of $X$ for predicting $Y$ and $Z$:

$$\rho(Y, Z|X) = \rho(Y - W, Z - U)$$

Verify this statement for the $X$, $Y$, and $Z$ of Problem 1.

3. Suppose $X$ and $Y$ are uncorrelated, that is, $\text{cov}(X,Y) = 0$. Simplify the formulas for $c$ and $d$ in this case. Do you recognize the result?

4. A coin with probability .6 of heads is tossed three times, and $X_i$ denotes the number of heads in the first $i$ tosses. Find the best linear function of:

(a) $X_1$ and $X_2$ for predicting $X_3$
(b) $X_1$ and $X_3$ for predicting $X_2$
(c) $X_2$ and $X_3$ for predicting $X_1$

5. For the scatter diagram of Figure 8.1, find the best linear function of $X$ and $X^2$ for predicting $Y$.

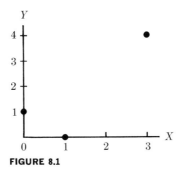

FIGURE 8.1

# 9 INDEPENDENCE

Urn 1

Urn 2

**FIGURE 9.1**

Two urns are shown in Figure 9.1. Two balls are drawn at random with replacement from urn 1. Then one ball is drawn at random from urn 2. The numbers on the balls drawn are denoted $X$, $Y$, and $Z$. It is clear that no one of the variables is helpful in predicting either of the other two, and even that no two of them are helpful in predicting the third. In fact, the values of any two tell us nothing at all about the third:

$$P(Y = 2 | X = 1, Z = 2) = P(Y = 2) = .25$$

We say that $X$, $Y$, and $Z$ are *independent*.

# INDEPENDENCE

Two or more variables $X, Y, \ldots$ are said to be *independent* if the distribution of each one, given the values of all the others, is always the same as its unconditional distribution, that is, does not depend on the values of the other variables.

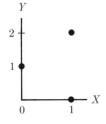

**FIGURE 9.2**

If two variables are independent, they are certainly uncorrelated. But Figure 9.2 shows a scatter diagram for two variables $X$ and $Y$ which are uncorrelated, but not independent.

The variance of a sum of independent variables is the sum of their variances:

$$\sigma^2(X + Y + \cdots) = \sigma^2(X) + \sigma^2(Y) + \cdots$$

if $X + Y + \cdots$ are independent.

More generally, the variance of a sum of variables is the sum of the variances whenever each two of the variables are uncorrelated. For instance, for two variables $X$ and $Y$,

$$\begin{aligned}\sigma^2(X + Y) &= \mathrm{cov}(X + Y, X + Y) \\ &= \mathrm{cov}(X, X + Y) + \mathrm{cov}(Y, X + Y) \\ &= \mathrm{cov}(X,X) + \mathrm{cov}(X,Y) + \mathrm{cov}(Y,X) \\ &\qquad + \mathrm{cov}(Y,Y) \\ &= \sigma^2(X) + \sigma^2(Y) + 2\,\mathrm{cov}(X,Y)\end{aligned}$$

Thus $\sigma^2(X + Y) = \sigma^2(X) + \sigma^2(Y)$ if $\mathrm{cov}(X,Y) = 0$, that is, if $X$ and $Y$ are uncorrelated.

An important case of independence is random sampling with replacement from a known distribution. For example, if ten balls are drawn at random with replacement from urn 1, and $X_1, \ldots, X_{10}$ are the numbers on the balls drawn, the variables $X_1, \ldots, X_{10}$ are independent with the same distribution:

| $v$ | $p$ |
|---|---|
| 1 | .5 |
| 2 | .25 |
| 3 | .25 |

The variables are said to be a *random sample* from this distribution.

On the other hand, if we selected one of the two urns at random, then drew ten balls at random with replacement from the selected urn, the numbers $X_1, \ldots, X_{10}$ on the balls drawn would not be independent. For if $X_1 = 3$, we know that urn 1 was selected, so $P(X_2 = 1|X_1 = 3) = \frac{1}{2}$, while the unconditional probability that $X_2 = 1$ is

$$P(X_2 = 1) = P(\text{urn 1 and } X_2 = 1) \\ + P(\text{urn 2 and } X_2 = 1)$$
$$= \tfrac{1}{2}(\tfrac{1}{2}) + \tfrac{1}{2}(\tfrac{1}{3}) = \tfrac{5}{12}$$

**EXAMPLE 9.1** A random sample $X_1, \ldots, X_{10}$ will be drawn from the following distribution:

| $v$ | $p$ |
|---|---|
| 1 | .4 |
| 2 | .2 |
| 3 | .4 |

You have to predict (a) the sum $S = X_1 + \cdots + X_{10}$ of the sample values, and (b) the average $(X_1 + \cdots + X_{10})/10 = S/10$ of the sample values, losing the square of your error. What is your best prediction, and what is your mse?

Your best prediction for $S$ is

$$E(S) = E(X_1) + \cdots + E(X_{10}) \\ = 2 + 2 + \cdots + 2 = 20$$

# INDEPENDENCE

and your mse is

$$\sigma^2(S) = \sigma^2(X_1) + \cdots + \sigma^2(X_{10})$$
$$= .8 + \cdots + .8 = 8$$

Your best prediction for $S/10$ is

$$E\left(\frac{S}{10}\right) = \frac{E(S)}{10} = \frac{20}{10} = 2$$

and your mse is

$$\sigma^2\left(\frac{S}{10}\right) = \frac{\sigma^2(S)}{10^2} = \frac{8}{100} = .08$$

Note that

$$\sigma\left(\frac{S}{10}\right) = \sqrt{.08}$$

$$\sigma(X_1) = \sqrt{.8}$$

So

$$\sigma\left(\frac{S}{10}\right) = \frac{\sigma(X_1)}{\sqrt{10}}$$

The variable $(X_1 + \cdots + X_{10})/10 = S/10$ is called the *sample mean*, and is often denoted by $\bar{X}$.

> If $X_1, \ldots, X_n$ are a random sample from a distribution, the mean of the sample mean is the mean $m$ of the distribution:
>
> $$E(\bar{X}) = E\left(\frac{X_1 + \cdots + X_n}{n}\right) = E(X_1)$$
> $$= E(X_n) = m$$
>
> and the standard deviation of the sample mean is the standard deviation $\sigma$ of the distribution divided by the square root of the size of the sample:
>
> $$\sigma(\bar{X}) = \frac{\sigma}{\sqrt{n}}$$

**BASIC STATISTICS**

**EXAMPLE 9.2**  A population is 80% male. We select a random sample with replacement of size 25, and $\bar{X}$ is the proportion of males in our sample. Find $E(\bar{X})$ and $\sigma(\bar{X})$.

If we write $X_i = 1$ if the $i$th individual is male, and $X_i = 0$ if the $i$th individual is female, then $X_1, \ldots, X_{25}$ are a random sample from the following distribution:

| $v$ | $p$ |
|---|---|
| 0 | .2 |
| 1 | .8 |

which has mean $m = 0(.2) + 1(.8) = .8$ and standard deviation

$$\sigma = \sqrt{.8 - (.8)^2} = \sqrt{.16} = .4$$

and

$$\bar{X} = (X_1 + \cdots + X_{25})/25$$

Thus

$$E(\bar{X}) = m = .8$$

$$\sigma(\bar{X}) = \frac{\sigma}{\sqrt{25}} = \frac{.4}{5} = .08$$

**DISCUSSION PROBLEM 9.1**  In Example 9.2, how large a sample would we have to take to have

$$\sigma(\bar{X}) = .04? \quad .02? \quad .01?$$

**PROBLEMS Set 9**

1. There are three urns. Each urn has one black ball, and urn $i$ has $i$ white balls. One ball is drawn at random from each urn, and you are paid $1 for each black ball drawn. Find the mean, variance, and standard deviation of your total income.

2. Balls are drawn at random with replacement from an urn containing ten balls labeled 0, 1, . . . , 9. You

are paid:

$1 when 0, 1, 2, or 3 is drawn
$2 when 4, 5, 6, 7, or 8 is drawn
$4 when 9 is drawn

Hence your income $Y_i$ from the $i$th draw is

$$Y_i = 1 \text{ for } X_i \leq 3$$
$$= 2 \text{ for } 4 \leq X_i \leq 8$$
$$= 4 \text{ for } X_i = 9$$

where $X_i$ is the label of the $i$th ball drawn.

Find the distribution of your income from the first two draws, and its mean and variance. Are $Y_1$, $Y_2$, . . . a sample from a distribution?

3. Suppose you had an urn like the one in Problem 2. How could you produce a random sample from the following distribution?

| $v$ | $p$ |
|---|---|
| 10 | .2 |
| 12 | .4 |
| 14 | .3 |
| 20 | .1 |

How could you produce a random sample from each of the following distributions?

| $v$ | $p$ |
|---|---|
| 10 | .18 |
| 12 | .43 |
| 14 | .28 |
| 20 | .11 |

| $v$ | $p$ |
|---|---|
| 0 | $\frac{1}{11}$ |
| 1 | $\frac{1}{11}$ |
| 2 | . |
| . | . |
| . | . |
| 10 | $\frac{1}{11}$ |

4. Experiments analogous to drawing balls at random with replacement from the urn in Problem 2 have been often carried out, and the results $X_1$, $X_2$, ... recorded in tables, called *random number tables*. Suppose you had a pair of fair dice of different colors, say one red and one green. How could you use the dice to construct a random number table? Could you use a single fair coin to construct a random number table? Construct in any way you please a random number table with 100 entries.

5. Suppose you have a coin whose probability of heads is .6. You toss the coin repeatedly, and pair the first two tosses, the next two, and so on. Then strike out all pairs with both tosses alike, and define $X_i = 1$ if the $i$th remaining pair is $HT$, or $X_i = 0$ if it is $TH$. If the first 12 tosses are

$$\cancel{HH} \quad TH \quad HT \quad \cancel{TT} \quad \cancel{TT} \quad HT$$

you get $\quad\quad\quad\quad\quad\; 0 \quad\quad 1 \quad\quad\quad\quad\quad\quad\; 1$
$\quad\quad\quad\quad\quad\quad\;\; X_1 \quad\; X_2 \quad\quad\quad\quad\quad\; X_3$

From what distribution, if any, are $X_1$, $X_2$, ... a random sample?

6. If $X$ and $Y$ are independent, $\mathrm{cov}(X,Y) = E(XY) - E(X)E(Y) = 0$, that is, $E(XY) = E(X)E(Y)$. More generally, if $X_1 X_2 \cdots X_n$ are independent,

$$E(X_1 X_2 \cdots X_n) = E(X_1)E(X_2) \cdots E(X_n)$$

Use this fact to find $E(X_1 X_2 \cdots X_n)$, where $X_1, \ldots, X_n$ are a random sample from the following distribution:

| $v$ | $p$ |
|---|---|
| 0 | .7 |
| 1 | .3 |

What is the distribution of $X_1 X_2 \cdots X_6$?

7. We continue random sampling from the following

distribution until we get a 1 or a 3:

| $v$ | $p$ |
|---|---|
| 1 | .4 |
| 2 | .3 |
| 3 | .2 |
| 4 | .1 |

Denote by $Y$ the final variable observed, and by $n$ the number of observations. Find $P(n = 1$ and $Y = 1)$, $P(Y = 1 \mid n = 1)$, $P(n = 2$ and $Y = 1)$, $P(Y = 1 \mid n = 2)$, $P(Y = 1)$, $P(n > 5)$, and $P(n \leq 5)$.

8. Figure 9.3 shows the three pages of a very short book. You select a page at random, then a line at random from the selected page, then a word at random from the

```
┌─────────────────┐  ┌─────────────────┐  ┌─────────────────┐
│               1 │  │               2 │  │               3 │
│ Excessive bail  │  │ sive fines imposed, │ ments inflicted. │
│ shall not be re │  │ nor cruel and un- │  │                 │
│ quired, nor exces- │ usual punish-  │  │                 │
└─────────────────┘  └─────────────────┘  └─────────────────┘
```

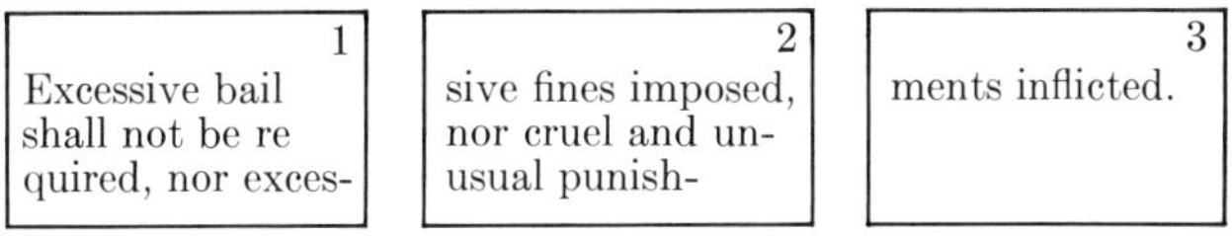

**FIGURE 9.3**

selected line. (Let's agree that a word belongs to the line on which it begins.) What is the probability that the word "shall" is selected? That "inflicted" is selected? Suppose you want to select a word at random from the book without first counting the number of words. You know only that (a) the book has three pages, (b) no page has more than five lines, and (c) no line has more than six words.

Here is a proposed method: Select $u = 1, 2, 3$ at random, $v = 1, 2, 3, 4, 5$ at random, $w = 1, 2, 3, 4, 5, 6$. Then look for the $w$th word on the $v$th line of the $u$th page. If there is such a word, that's it; if there isn't, select $u$, $v$, and $w$ again and repeat. Continue until you get a word. Thus (1,2,4) would produce "required," but (2,3,3) or (3,2,1) would not produce anything.

Explain why this process does or does not work.

# 10 BINOMIAL DISTRIBUTION

Smallpox vaccinations have been given to 30% of a population in the past five years. We select ten people at random from the population, with replacement. What is the chance that exactly four people in our sample have been vaccinated in the past five years? More generally, what is the distribution of $X$, the number of people in our sample who have been vaccinated in the past five years?

We imagine that we have a list of the $2^{10} = 1024$ possible results of our sampling experiment, as shown in the table at the top of the following page.

72

# BINOMIAL DISTRIBUTION

| Result | Probability | X |
|---|---|---|
| NNNNNNNNNN | $(.7)(.7) \ldots (.7)(.7) = (.7)^{10}$ | 0 |
| NNNNNNNNNY | $(.7)(.7) \ldots (.7)(.3) = (.3)(.7)^9$ | 1 |
| NNNNNNNNYN | $(.7)(.7) \ldots (.3)(.7) = (.3)(.7)^9$ | 1 |
| . . . . | | |
| YNNYNNNYY | $(.3)(.7) \ldots (.3)(.3) = (.3)^4(.7)^6$ | 4 |
| . . . . | | |
| YYYYYYYYYY | $(.3)(.3) \ldots (.3)(.3) = (.3)^{10}$ | 10 |

Opposite each result we have written its probability and the value of $X$, that is, the number of $Y$'s.

We can then find $P(X = 4)$ by going down the table, marking those results that have $X = 4$, and adding their probabilities. Each result with $X = 4$, that is, with 4 $Y$'s and 6 $N$'s, will have probability $(.3)^4(.7)^6$, so

$$P(X = 4) = C(10,4)(.3)^4(.7)^6$$

where $C(10,4)$ represents the number of listed words that have exactly 4 $Y$'s.

In general, we denote by $C(n,k)$ the number of words with $n$ letters, using only $N$ and $Y$ as letters, and having exactly $k$ $Y$'s. The following table shows values of $C(n,k)$ for $n = 1, 2, \ldots, 10$.

| $n$ \ $k$ | 0 | 1 | 2 | 3 | 4 | 5 | 6 | 7 | 8 | 9 | 10 |
|---|---|---|---|---|---|---|---|---|---|---|---|
| 1 | 1 | 1 | | | | | | | | | |
| 2 | 1 | 2 | 1 | | | | | | | | |
| 3 | 1 | 3 | 3 | 1 | | | | | | | |
| 4 | 1 | 4 | 6 | 4 | 1 | | | | | | |
| 5 | 1 | 5 | 10 | 10 | 5 | 1 | | | | | |
| 6 | 1 | 6 | 15 | 20 | 15 | 6 | 1 | | | | |
| 7 | 1 | 7 | 21 | 35 | 35 | 21 | 7 | 1 | | | |
| 8 | 1 | 8 | 28 | 56 | 70 | 56 | 28 | 8 | 1 | | |
| 9 | 1 | 9 | 36 | 84 | 126 | 126 | 84 | 36 | 9 | 1 | |
| 10 | 1 | 10 | 45 | 120 | 210 | 252 | 210 | 120 | 45 | 10 | 1 |

Can you explain the $k = 0$ column? The $k = 1$ column? The 1's down the diagonal? The fact that each row is symmetric about its center, for example, that $C(8,2) = C(8,6)$? What is the sum of the elements of the first row? The second row? The fifth row? What would be the sum of the elements of the eleventh row? Explain. Do you see a rule for calculating the numbers in one row from those in the row above? Use your rule to calculate the elements of the eleventh row. What is the sum of the elements you found? Can you explain why your rule works?

In our original problem,

$$P(X = 4) = 210(.3)^4(.7)^6$$

The distribution of $X$ is as follows:

| $v$ | $p$ |
|---|---|
| 0 | $(.7)^{10}$ |
| 1 | $10(.3)(.7)^9$ |
| 2 | $45(.3)^2(.7)^8$ |
| 3 | $120(.3)^3(.7)^7$ |
| 4 | $210(.3)^4(.7)^6$ |
| 5 | $252(.3)^5(.7)^5$ |
| 6 | $210(.3)^6(.7)^4$ |
| 7 | $120(.3)^7(.7)^3$ |
| 8 | $45(.3)^8(.7)^2$ |
| 9 | $10(.3)^9(.7)$ |
| 10 | $(.3)^{10}$ |

How would the distribution change if .3 were replaced by .31? If 10 were replaced by 8?

The number 10 in our problem is called the *sample size* or *number of trials*, the number .3 is called the *success probability* in a single trial, the variable $X$ is called the *number of successes in* 10 *independent trials with success probability* .3, and the distribution of $X$ is called the *binomial distribution with* 10 *trials and success probability* .3 or the *binomial distribution with* $n = 10$ *and* $p = .3$.

# BINOMIAL DISTRIBUTION

> If we draw a random sample of $n$ individuals with replacement from a population in which the probability of a certain attribute is $p$, the number $X$ of individuals in our sample with the attribute has $0, 1, \ldots, n$ as values, and
> $$P(X = k) = C(n,k)p^k q^{n-k}$$
> where $q = 1 - p$. The number $X$ is called a *binomial variable with parameters $n$ and $p$*, and its distribution is called a *binomial distribution with parameters $n$ and $p$*.

If $X_1, X_2, \ldots, X_n$ are a random sample from the distribution:

| $v$ | $p$ |
|---|---|
| 0 | $q$ |
| 1 | $p$ |

their sum $S = X_1 + X_2 + \cdots + X_n$ is a binomial variable with parameters $n$ and $p$, so

$$E(S) = p + p + \cdots + p = np$$
$$\sigma^2(S) = pq + pq + \cdots + pq = npq$$

> The mean and variance of a binomial variable $S$ with $n$ trials and success probability $p$ are $E(S) = np$ and $\sigma^2(S) = npq$, where $q = 1 - p$.

We now find a general formula for $C(n,k)$. To find $C(10,4)$, the number of ten-letter words using only $N$ and $Y$ as letters, containing exactly four $Y$'s, imagine an urn with ten balls, four labeled $Y$ and the other six labeled $N$. We draw the balls at random, without replacement, producing a ten-letter word. Here is one possible outcome:

*YNNYNNNYYN*

What is its probability?

$P(YNNYNNNYYN)$
$= \frac{4}{10} \times \frac{6}{9} \times \frac{5}{8} \times \frac{3}{7} \times \frac{4}{6} \times \frac{3}{5} \times \frac{2}{4} \times \frac{2}{3} \times \frac{1}{2} \times \frac{1}{1}$

Notice that the denominator is $10 \times 9 \times \cdots \times 3 \times 2 \times 1$. We denote this number by 10! (read "10 factorial"). The numerators for the $Y$'s are 4, 3, 2, 1, so their product is 4!, and the numerators for the $N$'s are 6, 5, 4, 3, 2, 1, so their product is 6! Thus our word has probability

$$\frac{4!(6!)}{10!} = \frac{4 \times 3 \times 2 \times 1}{10 \times 9 \times 8 \times 7} = \frac{1}{210}$$

For any other word we would again get

$$\frac{4!(6!)}{10!} = \frac{1}{210}$$

as its probability, so there must be

$$210 = \frac{10!}{4!(6!)} \text{ words}$$

$$C(10,4) = \frac{10!}{4!(6!)}$$

The same method gives the general formula:

$$C(n,k) = \frac{n!}{k!(n-k)!}$$

The formula will be correct for $k = 0$ and $k = n$ if we define 0! as 1.

PROBLEMS
Set 10

1. If $X$ is a binomial variable with parameters $n$ and $p$, find $P(X = k)$ for each set of values for $n$, $p$, and $k$

given in the following table:

| $n$ | 2 | 2 | 2 | 3 | 4 | 4 | 4 | 4 | 4 | 4 | 4 | 4 |
|---|---|---|---|---|---|---|---|---|---|---|---|---|
| $p$ | .4 | .4 | .4 | .4 | .4 | .4 | .5 | .6 | .3 | 2 | 0 | 1 |
| $k$ | 0 | 1 | 2 | 0 | 0 | 1 | 1 | 1 | 1 | 1 | 1 | 1 |
| $P(X = k)$ | | | | | | | | | | | | |

2. Draw the histogram for a binomial variable with $n = 5$ and each of these values for $p$: .4, .5, .6. What fraction of the area of the histogram is within one standard deviation of the mean?

3. A random sample of four individuals will be drawn with replacement from a certain population, and you win a prize if exactly one of the selected individuals has a certain attribute. Graph your probability of winning the prize as a function of $p$, the probability of the attribute in the population. What value of $p$ gives you the best chance? What is your chance for this value of $p$?

4. In a certain population 40% of the voters are Democrat. You can designate a sample size $n$, and that number of voters will be selected at random with replacement from the population. You win a prize if there are exactly two Democrats in the sample. What value of $n$ gives you the best chance? What is your chance for this value of $n$?

5. A variable $X$ is binomial with $n = 10$. Graph $E(X)$, $\sigma^2(X)$, $E\left(\dfrac{X}{10}\right)$, $\sigma^2\left(\dfrac{X}{10}\right)$ as functions of $p$, $0 \leq p \leq 1$.

6. A variable $X$ is binomial with $p = .4$. Graph $E(X)$, $\sigma^2(X)$, $E\left(\dfrac{X}{n}\right)$, $\sigma^2\left(\dfrac{X}{n}\right)$ as functions of $n$, $1 \leq n \leq 10$.

7. An urn has one black ball and $n - 1$ white balls. A selection of $n$ balls will be made at random from the urn, and you are paid $1 each time the black ball is

drawn. Find the mean and variance of your total income (a) with replacement, and (b) without replacement, for each of these values of $n$: 1, 2, 3, 5, 10. (c) Find the probability that your income is 0 for each of the above values of $n$, for drawing with replacement.

8. If $X$ is binomial with $n = 100$ and $p = .3$, express $P(X = 37)$ in terms of factorials. Express $P(X = 38)$. Find the ratio

$$r = \frac{P(X = 38)}{P(X = 37)}$$

Which is larger, $P(X = 38)$ or $P(X = 37)$? Find the ratio

$$\frac{P(X = k + 1)}{P(X = k)}$$

as a function of $k$. For what values of $k$ does this ratio exceed 1? For what values of $k$ is $P(X = k + 1)$ larger than $P(X = k)$? What is the most probable value of $X$?

9. Use the formula for $C(n,k)$ to check the table for $C(n,k)$ for $n = 2$ and for $n = 7$.

10. A fair coin is tossed $2k$ times, so that $X$, the number of heads, is a binomial variable with $n = 2k$ and $p = .5$, and the probability that the number of heads equals the number of tails is

$$p(k) = \frac{(2k)!}{(k!)(k!)} (.5)^k (.5)^k$$

Calculate $p(k)$ for $k = 1, 2, 3,$ and 4. Complete the table below, giving each entry to four significant figures:

| $k$ | $p(k)$ | $p^2(k)$ | $kp^2(k)$ | $(k + \tfrac{1}{2})p^2(k)$ |
|---|---|---|---|---|
| 1 | | | | |
| 2 | | | | |
| 3 | | | | |
| 4 | | | | |

Guess $p(5)$ from this table; then calculate it. About how large would $p(1000)$ be? $p(1,000,000)$? Do you think that the numbers $kp^2(k)$ increase as $k$ increases? That the numbers $(k + \frac{1}{2})p^2(k)$ decrease?

The rest of this problem requires algebra. Write a formula for the ratio

$$\frac{(k+1)p^2(k+1)}{kp^2(k)}$$

and simplify it as much as you can. Have you shown that the numbers $kp^2(k)$ increase? Write a formula for the ratio

$$\frac{(k+1+\frac{1}{2})p^2(k+1)}{(k+\frac{1}{2})p^2(k)}$$

and simplify it as much as you can. Have you shown that the numbers $(k + \frac{1}{2})p^2(k)$ decrease?

11. Each of two precincts has 100 voters. The number of Democrats in precinct $A$ is 20 and in precinct $B$ is 60. Following are three methods of selecting two voters:

(a) Select one at random from each precinct.
(b) Select two at random with replacement from the entire set of 200 voters.
(c) Select one of the two precincts at random, then select two voters at random with replacement from the selected precinct.

Find the mean, variance, and distribution of $S$, the number of Democrats selected, for each of the three methods given.

# 11 NORMAL APPROXIMATION

Following is the distribution of a binomial variable $X$ with $n = 6$ and $p = .4$, and Figure 11.1 shows its histogram.

| $v$ | $p$ |
|---|---|
| 0 | .047 |
| 1 | .187 |
| 2 | .311 |
| 3 | .276 |
| 4 | .138 |
| 5 | .037 |
| 6 | .004 |

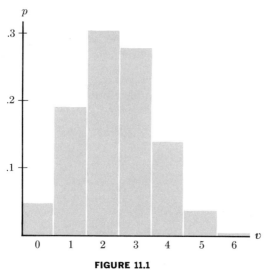

FIGURE 11.1

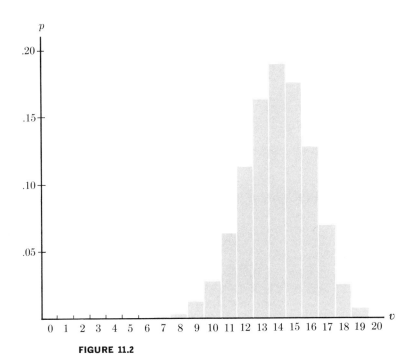

**FIGURE 11.2**

Figure 11.2 shows the histogram for a binomial variable with $p = .7$ and $n = 20$. The probabilities for $n = 20$ and for $n = 0, \ldots, 6$ are each less than .001.

It turns out that all binomial histograms for which $npq$ is large have about the same shape, that of a curve called the *normal curve*, shown in Figure 11.3. The total area under the curve is 1, and it is symmetric about 0.

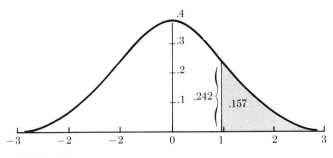

**FIGURE 11.3**

At the back of the book will be found a normal distribution table, showing the height $h(t)$ of the curve at $t$, and the area $H(t)$ under the curve to the right of $t$, for various values of $t$.

A variable with the normal curve as density is called a *standard normal variable*. It has mean 0 and standard deviation 1.

---

For any histogram, the normal approximation to the ratio

$$\frac{\text{Area under histogram between } a \text{ and } b}{\text{Total area under histogram}}$$

is the area under the normal curve between $t_a$ and $t_b$, where $t_x$ denotes the number of standard deviations $x$ is above the mean:

$$t_x = \frac{x - m}{\sigma}$$

where $m$ = mean of histogram
$\sigma$ = standard deviation of histogram

---

**EXAMPLE 11.1** $X$ is binomial with $n = 6$ and $p = .4$. Find the normal approximation to $P(3 \leq X \leq 5)$.

$P(3 \leq X \leq 5)$ is the part of the area under the histogram of $X$ between $a = 2.5$ and $b = 5.5$. Since $E(X) = 6(.4) = 2.4$ and $\sigma(X) = \sqrt{6(.4)(.6)} = 1.2$,

$$t_a = \frac{2.5 - 2.4}{1.2} = \frac{1}{12} = .083$$

$$t_b = \frac{5.5 - 2.4}{1.2} = \frac{31}{12} = 2.583$$

Thus the normal approximation to $P(3 \leq X \leq 5)$ is the area under the normal curve between $\frac{1}{12}$ and $\frac{31}{12}$, that is, $H(\frac{1}{12}) - H(\frac{31}{12})$. Interpolating between $H(.05) = .480$

and $H(.10) = .460$ gives $H(.083) = .467$. Similarly $H(2.583) = .0051$. So

$$P(3 \leq X \leq 5) = .467 - .0051 = .462$$

The correct value of $P(3 \leq X \leq 5)$ is

$$.276 + .138 + .037 = .451$$

Figure 11.4 shows a picture of the normal approximation to $P(3 \leq X \leq 5)$. The $(x,p)$ scale is for the binomial histogram, the $(t,h)$ scale is for the normal curve. Since $\sigma(X)$ is 1.2, so one $t$ unit is 1.2 $x$ units, we must make, on the vertical scale, one $p$ unit correspond to 1.2 $h$ units, to make the area units equal. Notice that, for each individual rectangle, the area under the corresponding part of the curve looks near that of the rectangle.

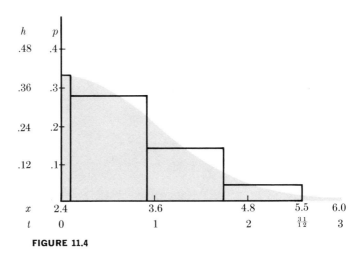

**FIGURE 11.4**

**EXAMPLE 11.2** A coin with probability .6 of heads is tossed 600 times. Find the normal approximation to the probability of exactly 372 heads.

We have

$$m = 600(.6) = 360$$
$$\sigma = \sqrt{600(.6)(.4)} = 12 \quad a = 371.5 \quad b = 372.5$$
$$t_a = \frac{371 - 5 - 360}{12} = \frac{115}{120}$$
$$t_b = \frac{372 - 5 - 360}{12} = \frac{125}{120}$$

Since $t_a$ and $t_b$ are close, we approximate the area $H(t_a) - H(t_b)$ by that of a rectangle with base $t_b - t_a = \frac{1}{12}$ and height $h[(t_a + t_b)/2] = h(1) = .242$ obtaining

$$P(372 \text{ heads}) = \tfrac{1}{12}(.242) = .202$$

The correct value to four places of $P(372 \text{ heads})$ is .0203.

**EXAMPLE 11.3** A sample $X_1, \ldots, X_{60}$ of size 60 is drawn from the following distribution:

| $v$ | $p$ |
|---|---|
| 0 | .3 |
| 1 | .4 |
| 2 | .3 |

Find the normal approximation to the probability that the sample average

$$\bar{X} = \frac{X_1 + \cdots + X_{60}}{60}$$

is below .9.

$$P(\bar{X} < .9) = P(X_1 + \cdots + X_{60} < 54)$$
$$E(X_1 + \cdots + X_{60}) = 60E(X_1) = 60(1) = 60$$
$$\sigma^2(X_1 + \cdots + X_{60}) = 60\sigma^2(X_1) = 60(.6) = 36$$
$$\sigma(X_1 + \cdots + X_{60}) = 6$$

Thus

$$P(X_1 + \cdots + X_{60} < 54) = 1 - H\left(\frac{53.5 - 60}{6}\right)$$
$$= 1 - H(-\tfrac{13}{12}) = H(\tfrac{13}{12}) = .140$$

There is about a 14% chance that the sample average will be below .9.

PROBLEMS
Set 11

1. A variable $X$ has mean 500, standard deviation 100, and is approximately normal. Find the probability of each of these events: $X \leq 400$, $400 \leq X \leq 650$, $X \geq 800$. Find the number $x$ for which $P(X > x) = .04$.

2. Draw a careful graph of $H(t)$ for $-2.5 \leq t \leq 2.5$. Could you use this graph instead of the normal distribution table to answer the questions in Problem 1?

3. If 100 balls are drawn at random with replacement from an urn in which 20% of the balls are black, what is the probability that at least 30 of the balls drawn are black?

4. An urn contains two red balls, one green ball, and one blue ball. If 1000 balls are drawn at random with replacement, what is the probability that at least 300 blues are obtained? That at least 550 reds are obtained?

5. If $X$ is a standard normal variable, find the probability that $X$ is in each of the intervals (0, .6), (.6, 1.2), (1.2, 1.8), (1.8, 2.4), and (2.4, 3.0), by (a) using the normal distribution table, or by (b) multiplying the value of $h$ at the midpoint of the interval by .6. Use either of these approximations to estimate $E(X^2)$, by approximating $X$ by a variable $X^*$ with the 10 values $\pm.3$, $\pm.9$, $\pm 1.5$, $\pm 2.1$, $\pm 2.7$.

6. A fair die is rolled 420 times. Find the normal approximation to the probability that the sum of the numbers produced is at least 1400.

7. If $X$ is binomial with nine trials and success probability .5, find the exact value of $P(X = 5)$, and the normal approximation.

8. A random sample with replacement of $n = 900$ people is drawn from a population in which the propor-

tion $p$ with a certain attribute is .7. Find the normal approximation to the probability that the proportion $\bar{X}$ in our sample is within $d = 2\%$ of the population proportion, or that $\bar{X}$ is between .68 and .72, or that $S$, the number in our sample with the characteristic, is between $(.68)(900) = 612$, and $(.72)(900) = 648$.

9. Repeat Problem 8 for each of the eight triples with $n = 225$ or $900$, $p = .5$ or $.7$, $d = 2\%$ or $5\%$:

| $n$ | $p$ | $d$ | $P(|\bar{X} - p| \leq d)$ |
|---|---|---|---|
| 225 | .5 | .02 | |
| 225 | .5 | .05 | |
| 225 | .7 | .02 | |
| 225 | .7 | .05 | |
| 900 | .5 | .02 | |
| 900 | .5 | .05 | |
| 900 | .7 | .02 | |
| 900 | .7 | .05 | |

10. A fair coin is tossed 400 times. (a) Find the normal approximation to the probability of exactly 200 heads. (b) Find upper and lower bounds on the exact probability from Problem 10 of Set 10.

11. If $X_1$ and $X_2$ are a sample from the distribution

| $v$ | $p$ |
|---|---|
| 0 | $p_0$ |
| 1 | $p_1$ |
| 2 | $p_2$ |
| . | . |
| . | . |
| . | . |
| $n$ | $p_n$ |

then $S = X_1 + X_2$ has the distribution

| $v$ | $p$ |
|---|---|
| 0 | $p_0^2$ |
| 1 | $p_0 p_1 + p_1 p_0$ |
| 2 | $p_0 p_2 + p_1 p_1 + p_2 p_0$ |
| 3 | $p_0 p_3 + p_1 p_2 + p_2 p_1 + p_3 p_0$ |
| . | |
| . | |
| . | |
| $n$ | $p_0 p_n + p_1 p_{n-1} + \cdots + p_n p_0$ |
| $n+1$ | $p_1 p_n + p_2 p_{n-1} + \cdots + p_n p_1$ |
| $n+2$ | $p_2 p_n + p_3 p_{n-1} + \cdots + p_n p_2$ |
| . | |
| . | |
| . | |
| $2n$ | $p_n^2$ |

(a) Explain $P(S = 3) = p_0 p_3 + p_1 p_2 + p_2 p_1 + p_3 p_0$.
(b) Find the distribution of $S$ when the distribution sampled is as follows:

| $v$ | $p$ |
|---|---|
| 0 | $\frac{1}{4}$ |
| 1 | $\frac{1}{4}$ |
| 2 | $\frac{1}{4}$ |
| 3 | $\frac{1}{4}$ |

(c) Find the distribution of $T = S_1 + S_2$, where $S_1$ and $S_2$ are a sample from the distribution of $S$ in (b). Why is the distribution of $T$ the same as the distribution of $X_1 + X_2 + X_3 + X_4$, where $X_1$, $X_2$, $X_3$, and $X_4$ are a sample from the distribution given in (b)? Draw the histogram for $T$. Find the normal approximation to $P(T = v)$ for each value $v$ of $T$.

# 12 INFERENCE

A pond has a small number $A$ of fish. You know that $A = 1, 2$, or $3$, and assign probabilities .2, .2, and .6 to these values. You catch a fish selected at random from the fish in the pond, tag it, and throw it back. The next day you again catch a fish selected at random, note whether it is the tagged fish or not, and throw it back. You must then decide whether the pond has three fish, winning a prize if your decision is correct. What is the best method of deciding, and how good are your chances of winning the prize?

Our first step in solving the problem is to represent what we are given in a table, which we shall call a *data table*:

**DATA TABLE**

| λ | A | X | |
|---|---|---|---|
| | | T | N |
| .2 | 1 | 1 | 0 |
| .2 | 2 | $\frac{1}{2}$ | $\frac{1}{2}$ |
| .6 | 3 | $\frac{1}{3}$ | $\frac{2}{3}$ |

The $A$ column lists the possible values of the variable $A$' and the $\lambda$ column shows the probability we assign to each value of $A$ before performing the experiment. The variable $A$, whose value is of interest to us, and which we hope to learn something about from the experiment, is called the *parameter* of the problem, and the distribution $\lambda$, which represents our opinion about $A$ before we perform the experiment, is called the *prior distribution* of the parameter. The variable $X$ represents the *result* or *outcome* of the experiment; our experiment has two possible results: $T$, the fish caught on the second day was the tagged fish, or $N$, it was not. The numbers in the $T$ column show the probability of result $T$ for each parameter value. If $A = 1$, we are certain to catch the tagged fish; if $A = 2$, we have probability $\frac{1}{2}$ of catching the tagged fish, and so on. The numbers in the $N$ column show the probability of result $N$ for each parameter value. The function $p(x|a)$, specifying for each outcome $x$ and parameter value $a$ the probability of getting outcome $x$ given that the parameter has value $a$, is called the *model* for the experiment. Thus our data table shows two things: (1) the prior distribution $\lambda$, and (2) the model $p$.

The second step is to construct a *joint distribution table*, showing for each parameter value $a$ and outcome $x$ the probability that the pair $(a,x)$ occurs. Since

$$P(A = a \text{ and } X = x)$$
$$= P(A = a)P(X = x|A = a) = \lambda(a)p(x|a)$$

we obtain the joint distribution table by multiplying each $p(x|a)$ in the data table by $\lambda(a)$:

**JOINT DISTRIBUTION TABLE**

| $A$ | $X$ | | $\Sigma$ |
|---|---|---|---|
|   | $T$ | $N$ |   |
| 1 | .2 | 0  | .2 |
| 2 | .1 | .1 | .2 |
| 3 | .2 | .4 | .6 |
| $\Sigma$ | .5 | .5 |   |

For instance,

$$P(A = 3 \text{ and } X = N) = P(A = 3)P(X = N|A = 3)$$
$$= (.6)(\tfrac{2}{3}) = .4$$

The $\Sigma$ row, which is the sum of the parameter rows, shows the (*overall* or *marginal*) distribution of $X$:

$$P(X = T) = .5 \text{ and } P(X = N) = .5$$

Similarly the $\Sigma$ column, the sum of the outcome columns, shows the distribution of $A$, which we already knew to be $\lambda$.

The third step is to construct a *posterior distribution table* showing, for each result $x$ of the experiment, the distribution we assign to $A$ when that result is observed. Since

$$P(A = a|X = x) = \frac{P(A = a \text{ and } X = x)}{P(X = x)}$$

we get the posterior distribution table by dividing each entry in the joint distribution table by the $\Sigma$ number at the bottom of its column:

**POSTERIOR DISTRIBUTION TABLE**

| $A$ | $X$ | |
|---|---|---|
| | $T$ | $N$ |
| 1 | .4 | 0 |
| 2 | .2 | .2 |
| 3 | .4 | .8 |

For instance,

$$P(A = 3|X = N) = \frac{P(A = 3 \text{ and } X = N)}{P(X = N)} = \frac{.4}{.5} = .8$$

Now we can solve our problem: Should you decide that $A = 3$? Since $P(A = 3|T) = .4 < .5$, your best decision if you catch the tagged fish is $A \neq 3$, and your chance of being right is .6. But if you catch an untagged fish, since $P(A = 3|N) = .8 > .5$, you should decide that $A = 3$, with an .8 chance of being correct.

Here are two ways of finding your overall chance of being correct:

Method 1

Of the six pairs $(a,x)$ that might occur, the ones that win the prize for you (using your decision procedure) are $(1,T)$, $(2,T)$, and $(3,N)$. The probabilities of these pairs, from the joint distribution table, are .2, .1, and .4, so your overall probability of being correct is

$$.2 + .1 + .4 = .7$$

Method 2

$$\begin{aligned}P(\text{prize}) &= P(T \text{ and prize}) + P(N \text{ and prize}) \\ &= P(T)P(\text{prize}|T) + P(N)P(\text{prize}|N) \\ &= .5(.6) + .5(.8) = .7\end{aligned}$$

Using your decision procedure, what is the probability that you win the prize if $A = 1$? If $A = 2$? If $A = 3$? (Answers: 1, $\frac{1}{2}$, $\frac{2}{3}$)

**Summary**

> If $T$ occurs, decide $A = 3$; $P(\text{prize}) = .6$
> If $N$ occurs, decide $A \neq 3$; $P(\text{prize}) = .8$
> Overall $P(\text{prize}) = .7$

Suppose that, instead of having to decide whether $A = 3$, you have to estimate $A$, losing the square of your error. What is the best estimate for $A$, and what is your mse? If $T$ occurs, your best estimate for $A$ is

$$E(A|T) = (.4)(1) + (.2)(2) + (.4)(3) = 2.0$$

and your mse is

$$\begin{aligned}\sigma^2(A|T) &= E(A^2|T) - [E(A|T)]^2 \\ &= (.4)(1)^2 + (.2)(2)^2 + (.4)(3)^2 - (2.0)^2 \\ &= 4.8 - 4.0 = .8\end{aligned}$$

If $N$ occurs, your best estimate for $A$ is $E(A|N) = 2.8$, and $\sigma^2(A|N) = 8 - (2.8)^2 = .16$.

Here are two ways of finding your overall mse:

**Method 1**

The following table shows the squared errors for the six $(a,x)$ pairs:

| $A$ | $X$ | |
|---|---|---|
| | $T$ | $N$ |
| 1 | $(1-2)^2 = 1$ | $(1-2.8)^2 = 3.24$ |
| 2 | $(2-2)^2 = 0$ | $(2-2.8)^2 = .64$ |
| 3 | $(3-2)^2 = 1$ | $(3-2.8)^2 = .04$ |

# INFERENCE

Multiply each squared error by the probability of the corresponding pair, getting

Overall mse $= 1(.2) + 0(.1) + 1(.2) + (3.24)(0)$
$\qquad + (.64)(.1) + (.04)(.4)$
$= .48$

**Method 2**

Overall mse $= P(T)(\text{mse}|T) + P(N)(\text{mse}|N)$
$= (.5)(.8) + (.5)(.16) = .48$

Your mse as a function of the number of fish in the pond is

$(\text{mse}|A = 1) = 1(1 - 2)^2 + D(1 - 2.8)^2 = 1$
$(\text{mse}|A = 2) = \frac{1}{2}(2 - 2)^2 + \frac{1}{2}(2 - 2.8)^2 = .32$
$(\text{mse}|A = 3) = \frac{1}{3}(3 - 2)^2 + \frac{2}{3}(3 - 2.8)^2 = .36$

**Summary**

If $T$ occurs, estimate $A = 2$; mse $= .8$
If $N$ occurs, estimate $A = 2.8$; mse $= .16$
Overall mse $= .48$

**Remarks**

1. If we have to decide whether $A = 3$ before fishing, our best decision is $A = 3$, and we have a .6 chance of being correct. The fishing experiment increases this probability to .7.

2. If we have to estimate $A$ before fishing, our best estimate is $E(A) = 2.4$ and our mse is $\sigma^2(A) = .64$. The fishing experiment reduces our mse to .48, so the best estimate from the fishing experiment has worth

$$\frac{.64 - .48}{.64} = .25$$

The outcome $T$ actually increases our mse to .8. We are more uncertain about $A$ after the experiment than we were before.

**PROBLEMS**
**Set 12**

1. In the data table shown below, you must observe $X$ and then decide which is the correct value of $A$, winning

a prize if you make the right choice. What is the best method, and what are your overall chances of winning the prize?

| λ | A | X | |
|---|---|---|---|
|   |   | B | W |
| .3 | a | .6 | .4 |
| .7 | b | .9 | .1 |

2. Describe an urn experiment with the data table of Problem 1.

3. Solve Problem 1 with the λ column replaced by:

(a) $\lambda(a) = .2$ and $\lambda(b) = .8$
(b) $\lambda(a) = .1$ and $\lambda(b) = .9$

4. The proportion $p$ of graduate students in a certain class is either .1, .2, or .5, and these proportions are equally likely. You will take a random sample of two individuals, with replacement, from the class, and record the status ($G$ or $U$) of each. Construct the data table for this experiment. Find the best estimate for $p$ when the outcome is $GU$, and find your mse for this outcome.

5. Suppose in Problem 4 someone does the sampling and tells you only the number $X$ of $G$'s in the sample. Construct your data table for this experiment, find the best estimate for $p$ when $X = 1$, and find your mse for $X = 1$.

6. Urn 1 has two black balls and one white ball; urn 2 has one black ball and two white balls. One of the urns is selected at random. Then two balls are drawn at random from the selected urn, and you will be told the number of black balls drawn. You must then decide which urn was selected. Make the data table for this problem

and find the best decision method and your overall chance of being correct (*a*) with replacement, or (*b*) without replacement.

7. One of the three words in the sentence THIS IS IT is selected at random. Then a letter is selected at random from the selected word and shown to you. Then you must guess which word was selected, winning a prize if you are correct. Make the data, joint distribution, and posterior distribution tables. Find the best decision method, your chance for each outcome of winning the prize, and your overall chance of winning the prize. Using your method, what is your chance of winning the prize if the selected word is THIS? IS? IT?

8. A point is selected at random from the scatter diagram of Figure 12.1. You observe $X$ and must estimate

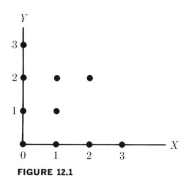

**FIGURE 12.1**

$Y$, losing the square of your error. Make the data, joint distribution, and posterior distribution tables, and find the best decision method and your overall mse.

9. An urn contains $n$ balls numbered $1, 2, \ldots, n$. You know that $n = 3$, 4, or 5, and consider these values equally likely. You draw a ball at random from the urn. What is the posterior distribution of $n$ if you draw ball number 1? 2? 3? 4? 5?

# 13 INFERENCE ABOUT PROPORTIONS (I)

Suppose you have to decide whether the proportion of smokers in a certain population exceeds 25%. You take a random sample of 100 people, and find 30 smokers. What should you decide, and what is the chance that you are correct?

The parameter of interest $A$, the proportion of smokers in the population, has a large number of possible values: If there are 10,000 members of the population, any of the 10,001 numbers $0, \frac{1}{10,000}, \frac{2}{10,000}, \ldots, 1$ is a possible value. Further, we often won't know the exact size of the population. We imagine that *all* numbers between 0 and 1 are possible values for $A$ and describe our prior opinion about $A$ by a density function $f$, defined for $0 \leq x \leq 1$, called the *prior density* of $A$.

Any density on (0,1) is a possible prior density. We looked at several densities on (0,1) in Chapter 5.

The relation between the prior density of a proportion and its posterior density, given a sample, is very simple:

> If $f$ is the prior density of a proportion, and a random sample yields $u$ individuals with the characteristic and $v$ without it, the posterior density of the proportion is
>
> $$g(x) = f(x)x^u(1-x)^v$$

For instance, in our smoking example, if our prior opinion for $A$ is uniform on $(0,\frac{1}{2})$, that is $f(x) = 1$ for $0 \leq x \leq \frac{1}{2}$, and $f(x) = 0$ for $x > \frac{1}{2}$, our posterior distribution for $A$ has density

$$g(x) = \begin{cases} x^{30}(1-x)^{70} & \text{for } 0 \leq x \leq \frac{1}{2} \\ 0 & \text{for } \frac{1}{2} < x \leq 1 \end{cases}$$

We have to draw the graph of $g$ and find the proportion of the total area under $g$ that is between 0 and .25. This is beyond us, but we can deal with very small samples. For instance, if we select one individual and he is a nonsmoker, we have $u = 0$ and $v = 1$, so the posterior density of $A$ is

$$g(x) = \begin{cases} (1-x) & \text{for } 0 \leq x \leq \frac{1}{2} \\ 0 & \text{for } \frac{1}{2} < x \leq 1 \end{cases}$$

As seen in Figure 13.1, the total area under $g$ is

$$\frac{1}{2}\left(\frac{1+\frac{1}{2}}{2}\right) = \frac{3}{8}$$

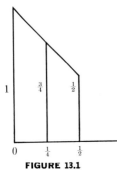

**FIGURE 13.1**

and the area under $g$ between 0 and $\frac{1}{4}$ is

$$\frac{1}{4}\left(\frac{1+\frac{3}{4}}{2}\right) = \frac{7}{32}$$

so $P(0 \leq A \leq \frac{1}{4}|\text{sample}) = \frac{7}{32}/\frac{3}{8} = \frac{7}{12} = .57$. The probability that $A \leq \frac{1}{4}$ has increased from .5 to .57.

**DISCUSSION PROBLEM 13.1**

Find $P(0 \leq A \leq \frac{1}{4}|\text{sample})$ if our individual is a smoker. (Answer: $\frac{1}{4}$)

If we took a sample of five people, and found two smokers, so that $u = 2$ and $v = 3$, we would have

$$g(x) = \begin{cases} x^2(1-x)^3 & \text{for } 0 \leq x \leq \frac{1}{2} \\ 0 & \text{for } x > \frac{1}{2} \end{cases}$$

We calculate values for $g$:

| $x$ | $x^2$ | $(1-x)^3$ | $g$ |
|---|---|---|---|
| 0 | 0 | 1 | 0 |
| .1 | .01 | .729 | .00729 |
| .2 | .04 | .512 | .02048 |
| .3 | .09 | .343 | .03087 |
| .4 | .16 | .216 | .03456 |
| .5 | .25 | .125 | .03125 |

Draw a graph for $g$ (or any convenient multiple, say $1000g$), and estimate the areas under $g$ in $(0,\frac{1}{4})$ and $(\frac{1}{4},\frac{1}{2})$. As seen in Figure 13.2, the area between 0 and $\frac{1}{4}$ looks about equal to the area under the dotted line at height 12, and the area between $\frac{1}{4}$ and $\frac{1}{2}$ looks about equal to the area under the dotted line at height 32, so we estimate

$$P(A \geq .25|\text{sample}) = \frac{32(.25)}{32(.25) + 12(.25)} = \frac{8}{11} = .73$$

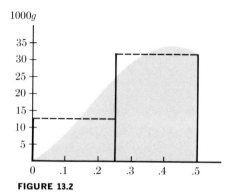

**FIGURE 13.2**

If our sample had had all smokers, that is, if $u = 5$ and $v = 0$, we would have, as graphed in Figure 13.3,

$$g(x) = \begin{cases} x^5 & \text{for } 0 \leq x \leq \tfrac{1}{2} \\ 0 & \text{for } x > \tfrac{1}{2} \end{cases}$$

The area between $\tfrac{1}{4}$ and $\tfrac{1}{2}$ is about $12(.25) = 3$. The area between 0 and $\tfrac{1}{4}$ is too small to be estimated from the

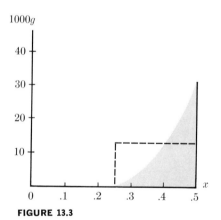

**FIGURE 13.3**

graph. Let's ignore the really negligible area below .2, and calculate

$1000g(.2) = 1000(.2)^5 = .32$
$1000g(.25) = 1000(.25)^5 = .98$

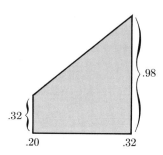

**FIGURE 13.4**

and estimate the area between .2 and .25 as the area of the trapezoid in Figure 13.4, which is $(.05)\left(\dfrac{.32 + .98}{2}\right) = .0325$.

Thus we estimate

$$P(A \geq .25 | \text{sample}) = 3/(3.0325) = .99$$

Even a sample of five can yield very strong evidence. (The true value of $P(A \geq .25 | \text{sample})$ is $\frac{63}{64} = .985$.)

**PROBLEMS**
**Set 13**

1. For prior density $f(x) = x^2(1 - x)$ and a sample with $u = 1$ and $v = 2$, graph the posterior density of $A$, and estimate $\sigma^2(A | \text{sample})$.

2. The area under the graph of $x^n$ between 0 and $t$ is $\dfrac{t^{n+1}}{n+1}$. Check this fact for the following values: (a) $t = \frac{1}{2}$, $n = 0$, (b) $t = \frac{1}{2}$, $n = 1$, (c) $t = \frac{1}{2}$, $n = 2$ (approximately).

3. Use the fact given in Problem 2 to show that if $A$ has a uniform prior distribution on $(0, \frac{1}{2})$ and a sample of five has all successes, then

$$P(A \geq .25 | \text{sample}) = \tfrac{63}{64}$$

4. A sample of three has all successes. If $A$ has a uniform prior distribution, use the fact given in Problem 2

to find the numbers
$$P_1 = P(0 \le A \le .4|\text{sample})$$
$$P_2 = P(.4 \le A \le .8|\text{sample})$$
$$P_3 = P(.8 \le A \le 1|\text{sample})$$

Find the mean of the approximating variable $A^*$ with the following distribution:

| $v$ | $p$ |
|---|---|
| .2 | $P_1$ |
| .6 | $P_2$ |
| .9 | $P_3$ |

5. A proportion $A$ has one of the 11 values $0, .1, .2, \ldots, .8, .9, 1$, and these values are a priori equally likely. A random sample from the population yields $u$ successes and $v$ failures. Draw the posterior histogram for $A$, given the following samples:

(a) $u = 1, v = 0$
(b) $u = 1, v = 1$
(c) $u = 2, v = 0$

# 14 INFERENCE ABOUT PROPORTIONS (II)

If you start with a uniform prior density on (0,1) for a proportion $A$, and a random sample yields $u$ individuals with a certain characteristic and $v$ without it, the resulting posterior density for $A$ is

$$x^u(1-x)^v$$

We shall call this density *the $(u,v)$ density*.

Figure 14.1 shows the graphs of some $(u,v)$ densities. The top three all have $u = v$. They are symmetric about .5, but get narrower as $u$ increases: The larger the sample

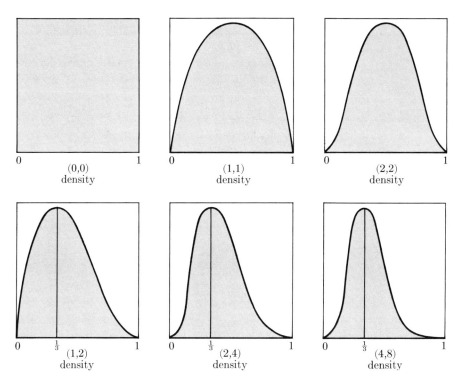

**FIGURE 14.1**

we have with $u = v$, the more likely it is that $A$ is near .5. The bottom three all have $v = 2u$: Twice as many failures as successes. They all have a maximum at $\frac{1}{3}$, and get narrower as $u$ increases: The larger the sample we have with $v = 2u$, the more likely it is that $A$ is near $\frac{1}{3}$.

**EXAMPLE 14.1** Suppose your prior density on a proportion $A$ is uniform, that is, the (0,0) density. You take a sample of 100 and get 30 successes. What is your best estimate for $A$? What is the probability $P$ that the true value of $A$ is within 5% of your estimated value?

Since your posterior density is the (30,70) density, your best estimate is the mean $m$ of the (30,70) density, and $P$ is the probability that a variable with a $(30,70)$ density is within .05 of its mean.

For such a $(u,v)$ density, the normal approximation will be close, so all we need is the mean and variance of the density.

If $A$ has a $(u,v)$ density,

$$E(A) = \frac{u+1}{u+v+2}$$

$$\sigma^2(A) = \frac{(u+1)(v+1)}{(u+v+2)^2(u+v+3)}$$

For large $u$ and $v$, simpler approximate formulas are

$$E(A) = \frac{u}{u+v} = p$$

$$\sigma^2(A) = \frac{uv}{(u+v)^3} = \frac{pq}{n}$$

where $p = u/(u+v)$, $q = 1-p$, and $n = u+v$.

Thus if $A$ has a $(30,70)$ density, approximately

$$E(A) = \frac{30}{30+70} = .3$$

$$\sigma^2(A) = \frac{(.3)(.7)}{100} = .0021$$

$$\sigma(A) = \sqrt{.0021} = .046$$

$P(A \text{ is within } .05 \text{ of } .3) = P(.25 \leq A \leq .35)$

$$= H\left(\frac{.25-.3}{.046}\right) - H\left(\frac{.35-.3}{.046}\right)$$

$$= H(-1.09) - H(1.09) = 1 - 2H(1.09)$$

$$= 1 - 2(.138) = .724$$

Your best estimate for $A$ is .30, the success proportion in the sample. Your mse is $\sigma^2(A) = .0021$, and the chance that you are within 5% of the correct value of $A$ is about 72%.

How much difference does it make what prior opinion you start with? In our example, if you had started with the (4,1) density, your posterior density after the (30,70) sample would be

$$g(x) = x^4(1-x)x^{30}(1-x)^{70} = x^{34}(1-x)^{71}$$

or, in other words, the (34,71) density. Your estimate for $A$ would be

$$E(A) = \frac{34}{105} = .324$$

and

$$\sigma^2(A) = \frac{(.324)(.676)}{105} = .0021 \qquad \sigma(A) = .046$$

$P(A \text{ is within } .05 \text{ of } .324) = .724$

This example illustrates two facts:

1. If your prior opinion is rather well spread out and you have a reasonably large sample, your posterior opinion will be almost independent of your prior opinion.

2. If your prior opinion has the $(r,s)$ density and you observe a $(u,v)$ sample, your posterior opinion has the $(r+u, s+v)$ density.

The second of these two facts has a very natural interpretation: If you start with a uniform prior opinion about $A$, that is, with the (0,0) density, and get an $(r,s)$ sample, your posterior opinion then has the $(r,s)$ density. If you now take a second sample and observe $(u,v)$, your final opinion can be calculated either as the result of a $(u,v)$ sample on an $(r,s)$ prior opinion, or (pooling the two samples) the result of an $(r+u, s+v)$ sample on a (0,0) prior opinion. The second fact says that the two methods of calculation yield the same result.

**EXAMPLE 14.2** A man is rather strongly convinced that the proportion $A$ of male births in a population is nearly .5, and assigns the (500,500) distribution to $A$. What will be his estimate for $A$ $(a)$ after a sample of 100 births yields 52 males? $(b)$ After a sample of 10,000 births yields 5200 males?

For $(a)$:
$$E(A) = \frac{500 + 52}{(500 + 52) + (500 + 48)} = \frac{552}{1100} = .502$$

For $(b)$:
$$E(A) = \frac{500 + 5200}{(500 + 5200) + (500 + 4800)} = \frac{5700}{11000} = .518$$

After a sample of 100, his prior opinion still dominates, but a sample of 10,000 overwhelms his prior opinion.

## PROBLEMS Set 14

1.$(a)$ Find a $(u,v)$ density that approximates your opinion about the proportion $A$ of students on your campus whose surnames begin with $A$ through $F$.

$(b)$ Select some number of students, between 20 and 100, say, as nearly at random as you can easily manage, and record the number $x$ of students in your sample whose names begin with $A$ through $F$ and the number $y$ whose names begin with $G$ through $Z$. One possibility is to take an issue of the campus newspaper and treat the students whose names appear in that issue as a random sample.

$(c)$ Find the mean $m$ and standard deviation $\sigma$ of your posterior distribution for $A$.

$(d)$ Now estimate the value of $A$ as accurately as you can from the student directory. Where does this value of $A$ fall in your posterior distribution, or what is $(A - m)/\sigma$?

2. Repeat Problem 1, but replace $A$ through $F$ by $A$ through $L$, students by residents of your city, the campus newspaper by a local city newspaper (sampling rather than complete listing), and the student directory by the telephone book.

3. Consider the population whose elements are sequences of five digits: 00000, 00001, . . . , 99999.

Among the 100,000 members of this population, a certain proportion $A$ have the property that all five digits are different. For instance, 21387 has this property, but 10207 does not. Start with a uniform prior opinion about $A$, and use a table of random numbers to select a sample from this population, continuing until your posterior distribution has $\sigma \leq .05$, so that you are about 68% sure that the true value of $A$ is within .05 of your estimated value $(H(-1) - H(1) = .68)$. Now calculate the exact value of $A$.

4. Estimate (by sampling) the proportion $A$ of left-handed students on your campus. Find the value of $P(A \geq .2|\text{sample})$.

# 15 INDEPENDENT PROPORTIONS

If $A_1$ and $A_2$ are two independent proportions, that is, knowledge of $A_1$ tells us nothing about $A_2$, and $c_1$ and $c_2$ are any constants, then

$$X = c_1 A_1 + c_2 A_2$$

has mean and variance

$$m = c_1 m_1 + c_2 m_2$$
$$\sigma^2 = c_1^2 \sigma_1^2 + c_2^2 \sigma_2^2$$

where $m_i$ and $\sigma_i^2$ are the mean and variance of $A_i$.

**INDEPENDENT PROPORTIONS**

Two interesting cases are $c_1 = 1$ and $c_2 = -1$, so $X = A_1 - A_2$ is the *difference between two proportions*; and $c_1 \geq 0$, $c_2 \geq 0$, $c_1 + c_2 = 1$, so $X$ is the *weighted average of two proportions*.

**EXAMPLE 15.1**  Two drugs are to be tested for effectiveness in treating hylosis. We are interested in the proportion $A_i$ of hylotics who improve when given drug $i$; before experimentation $A_1$ and $A_2$ are considered independent and uniform on $(0,1)$. We select 100 hylotics at random and give them drug 1; 20 of them improve. We select 300 hylotics at random and give them drug 2; 45 of them improve. What is the chance that drug 1 is better than drug 2, or that $A_1 - A_2 > 0$? From our data,

$$m_1 = \frac{20}{100} = .2 \qquad \sigma_1{}^2 = \frac{(.2)(.8)}{100} = .0016$$

$$m_2 = \frac{45}{300} = .15 \qquad \sigma_2{}^2 = \frac{(.15)(.85)}{300} = .000425$$

Thus

$$X = A_1 - A_2$$

has

$$m = .2 - .15 = .05$$
$$\sigma^2 = .0016 + .000425 = .002025$$
$$\sigma = .045$$

The normal approximation gives

$$P(X > 0) = H\left(\frac{0 - .05}{.045}\right) = 1 - H(1.11) = 1 - .134$$
$$= .866$$

There is about an 87% chance that drug 1 is really better than drug 2.

**EXAMPLE 15.2**  We want to estimate the proportion $A$ of smokers in a population; we know that 30% of the population is female and consider the proportions $A_1$ and $A_2$ of smokers among females and males as independent with (0,0) prior densities. A random sample contains 300 males, of whom 120 are smokers, and 200 females, of whom 100 are smokers. Find the mean and standard deviation of the proportion $A$ of smokers in the population.

Here is the result of our sample in a $2 \times 2$ table:

|         | Smokers | Nonsmokers |     |
| ------- | ------- | ---------- | --- |
| Male    | 120     | 180        | 300 |
| Female  | 100     | 100        | 200 |
|         | 220     | 280        |     |

$$m_1 = \frac{100}{200} = .5 \qquad \sigma_1^2 = \frac{(.5)(.5)}{200} = .00125$$

$$m_2 = \frac{120}{300} = .4 \qquad \sigma_2^2 = \frac{(.4)(.6)}{300} = .0008$$

Thus, since $A = .3A_1 + .7A_2$, the mean and standard deviation of $A$ are

$$m = (.3)(.5) + (.7)(.4) = .43$$
$$\sigma^2 = (.3)^2(.00125) + (.7)^2(.0008) = .0005045$$
$$\sigma = \sqrt{.0005045} = .0225$$

Suppose that, in the drug example, we had given each drug to 200 patients and obtained the same success ratios as before: 20% improve on drug 1 and 15% on drug 2. We then have

$$m_1 = .2 \qquad \sigma_1^2 = \frac{(.2)(.5)}{200} = .0008$$

$$m_2 = .15 \qquad \sigma_2^2 = \frac{(.15)(.85)}{200} = .000638$$

So

$$X = A_1 - A_2$$

## INDEPENDENT PROPORTIONS

has

$m = .2 - .15 = .05$     as before
$\sigma^2 = .0008 + .000638 = .001438$
$\sigma = \sqrt{.001438} = .038$     smaller than before

Obviously, our precision is somewhat better from the (200,200) allocation than from the (100,300) allocation.

In the smoking example, if we had selected 250 females and 250 males instead of 200 and 300, and obtained the same $m_1 = .5$ and $m_2 = .4$, we would have found

$m = (.3 \times .5) + (.7 \times .4) = .43$     as before
$\sigma^2 = (.3)^2 \dfrac{(.5)(.5)}{250} + (.7)^2 \dfrac{(.4)(.6)}{250} = .0056$
$\sigma = .024$     larger than before

Our precision is somewhat worse from the equal (250,250) allocation than from the (200,300) allocation.

For a given total sample size, what allocations to $A_1$ and $A_2$ give the most precision for $c_1 A_1 + c_2 A_2$?

> For given sample proportions $m_1$ and $m_2$, and fixed total sample size, the most precision in $c_1 A_1 + c_2 A_2$, that is, the smallest standard deviation, is obtained when the sample size on $A_i$ is proportional to
>
> $|c_i| \sqrt{m_i(1 - m_i)}$

The usefulness of this statement is somewhat restricted by the fact that we have to decide on allocation before sampling, or before $m_1$ and $m_2$ are known. A common procedure is to base the allocation on rough estimates of $m_1$ and $m_2$, obtained from small preliminary sampling, for instance.

Thus, in our drug example, if we had estimated $m_1 = m_2 = .05$ before sampling, the allocation of patients to drug 1 and drug 2 would be in the ratio

$$\frac{1\sqrt{(.05)(.95)}}{1\sqrt{(.05)(.95)}} \quad c_1 = 1 \text{ and } c_2 = -1$$

that is, equal allocation. If we had been lucky enough to estimate $m_1 = .2$ and $m_2 = .15$ in advance, our allocation would have been in the ratio

$$\frac{1\sqrt{(.2)(.8)}}{1\sqrt{(.15)(.85)}} = 1.12$$

that is, 112 patients on drug 1 to each 100 on drug 2, so

$\frac{112}{212}(400) = 211$ patients on drug 1

$\frac{100}{212}(400) = 189$ patients on drug 2

Similarly, in the smoking example, the best allocation for $m_1 = .5$ and $m_2 = .4$ is

$$\frac{.3\sqrt{(.5)(.5)}}{.7\sqrt{(.4)(.6)}} = .44$$

or about 44 females for each 100 males, so the sample would have

$\frac{100}{144}(500) = 347$ males

$\frac{44}{144}(500) = 153$ females

**PROBLEMS Set 15**

1. We shall call a word "long" if it contains more than six letters. Take two novels, by different authors, and denote by $A_1$ and $A_2$ the proportions of long words in the novels. Start with independent uniform prior distributions for $A_1$ and $A_2$, select 50 words at random from each novel, find the posterior mean and standard deviation for $A_1$ and $A_2$ and find $P(A_1 > A_2)$.

Now assume that the number of words in each novel is proportional to the number of pages in it, and find the posterior mean and standard deviation for $A$, the proportion of long words in the two novels together.

2. Get a solid cylinder whose height is approximately two-thirds its diameter, for example by cutting off a piece of a broom handle or taping eight pennies together. What is your prior opinion about the probability $A$ that this cylinder will come to rest on its side when tossed? Decide whether $A \geq \frac{1}{2}$ by tossing the cylinder until you are 90% certain that your answer is correct.

3. Get a thumbtack and decide whether the probability $B$ that the thumbtack comes to rest on its side when tossed is $\geq \frac{1}{2}$, by tossing it until you are 90% certain that your answer is correct.

4. Decide whether $A \geq B$, using the data from Problems 2 and 3, and making additional tosses of the cylinder and/or the thumbtack if necessary.

# 16 CHI SQUARE

We take a sample of 100 individuals, and classify each person by sex ($M$ or $F$) and by hair color ($D$ for dark or $L$ for light), obtaining a 2 × 2 table:

|   | $D$ | $L$ |    |
|---|-----|-----|----|
| $M$ | 20 | 40 | 60 |
| $F$ | 10 | 30 | 40 |
|   | 30 | 70 |    |

If there were no association in the population between sex and hair color we would expect, since 30% of the sample has dark hair, that 30% of the males in the sample, or 18, would have dark hair. Our *expected* 2 × 2 table is as follows:

|   | D | L |    |
|---|---|---|----|
| M | 18 | 42 | 60 |
| F | 12 | 28 | 40 |
|   | 30 | 70 |    |

A common measure of the discrepancy between the two tables, that is, how far the observed table differs from what would be expected under no association, is the $x^2$ (chi square) statistic:

$$x^2 = \text{sum over cells of } \frac{(\text{observed} - \text{expected})^2}{\text{expected}}$$

For our table,

$$x^2 = \frac{(20-18)^2}{18} + \frac{(40-42)^2}{42}$$
$$+ \frac{(10-12)^2}{12} + \frac{(30-28)^2}{28}$$
$$= .222 + .095 + .333 + .143$$
$$= .793$$

The probability, if there is no association, of getting a $x^2$ at least as large as the one observed is called the *level* of the observed $x^2$, and is given approximately by the following formula:

Level of $x^2 = 2H(\sqrt{x^2})$

Thus the level of our $x^2$ is

$$2H(\sqrt{.793}) = 2H(.89) = 2(.187) = .374$$

If there were no association between sex and hair color in the population, about 37% of the time we would get a sample at least as far from the expected, as measured by $\chi^2$, as the one we did get. Assume that our observed table had been as follows:

|   | D  | L  |    |
|---|----|----|----|
| M | 30 | 30 | 60 |
| F | 10 | 30 | 40 |
|   | 40 | 60 |    |

Then the expected table under no association would be the following:

|   | D  | L  |    |
|---|----|----|----|
| M | 24 | 36 | 60 |
| F | 16 | 24 | 40 |
|   | 40 | 60 |    |

So

$$\chi^2 = \frac{(30-24)^2}{24} + \frac{(30-36)^2}{36} + \frac{(10-16)^2}{16} + \frac{(30-24)^2}{24}$$

$$= 1.5 + 1 + 2.25 + 1.5 = 6.25$$

The level of our $\chi^2$ would then be

$$2H(\sqrt{6.25}) = 2H(2.5) = 2(.00621) = .0124$$

Only once in 100 samples, under no association, would we get a sample as extreme as this one.

The level does *not* measure the posterior probability that the two characteristics are not associated. But half the level, namely $H(\sqrt{x^2})$, does measure, roughly, the posterior probability that the difference in population proportions has the sign ($\pm$) opposite to that in our sample, starting with the prior opinion that the two proportions are independent and uniform (so that equality is impossible). For instance, if $A_1$ and $A_2$ are the proportions of $D$'s in the $(M,F)$ populations, our first observed table has

$$m_1 = \tfrac{1}{3} \quad m_2 = \tfrac{1}{4}$$

$$\sigma_1^2 = \frac{\tfrac{1}{3}(\tfrac{2}{3})}{60} = .00370$$

$$\sigma_2^2 = \frac{\tfrac{1}{4}(\tfrac{3}{4})}{40} = .00469$$

So $A_1 - A_2$ has $m = \tfrac{1}{3} - \tfrac{1}{4} = \tfrac{1}{12} = .0833$.

$$\sigma^2 = .00470 + .00469 = .00839$$

$$\sigma = \sqrt{.00839} = .092$$

$$P((A_1 - A_2) < 0) = H\left(\frac{.0833}{.092}\right) = H(.905) = .183$$

as compared with the half-level .187.

For the second observed table,

$$m_1 = .5 \quad m_2 = .25$$

$$\sigma_1^2 = \frac{(.5)(.5)}{60} = .00417$$

$$\sigma_2^2 = .00469$$

$$m = .25$$

$$\sigma = \sqrt{.00417 + .00469} = \sqrt{.00886} = .094$$

$$P((A_1 - A_2) < 0) = H\left(\frac{.25}{.094}\right) = H(2.66) = .0039$$

The half-level of the $x^2$ is .00621.

If the hypothesis of equality of two proportions is taken seriously (and it rarely is), the posterior probability of equality is approximately

$$Q = \frac{PR}{PR + 1 - P}$$

where $P$ is the prior probability of equality and

$$R = \frac{h\left(\frac{m}{\sigma}\right)}{\sigma}$$

If $R = 1$, we have $Q = P$ and the sample gives no evidence for or against equality. If $R > 1$, we have $Q > P$ and the sample is evidence for equality. If $R < 1$, we have $Q < P$ and the sample is evidence against equality. For instance, with our first table,

$$R = \frac{h(.905)}{.092} = \frac{.265}{.092} = 2.88$$

The sample is evidence for no association between sex and hair color, that is, evidence for equality of the proportions of males and females with dark hair. For instance, if $P = .5$, that is, association and no association are a priori equally likely, then

$$Q = \frac{(.5)(2.88)}{(.5)(2.88) + .5} = .74$$

so the posterior probability of no association is .74.
If $P = .1$, then

$$Q = \frac{(.1)(2.88)}{(.1)(2.88) + .9} = .24$$

Association is still less likely than no association, but more likely than it was before sampling.

With our second table,

$$R = \frac{h\left(\frac{.25}{.094}\right)}{.094} = \frac{h(2.64)}{.094} = .13$$

so the sample is substantial evidence against no association. For example, if $P = .5$, then

$$Q = \frac{(.5)(1.3)}{(.5)(.13) + .5} = .115$$

**PROBLEMS Set 16**

1. We take a sample of $2n$ from a population, obtaining the following table:

|   | D | L |
|---|---|---|
| M | .5n | .5n |
| F | .4n | .6n |

Calculate $x^2$, its level, and $R$ for $n = 100$ and for $n = 400$, and interpret the results.

2. Are words that contain at least one $E$ ($E$-words) more likely also to be $T$-words (to contain at least one $T$) than are not-$E$-words? Answer this question for the population of words in some book, by random sampling, continuing to sample until you are 90% sure that your answer is correct.

3. Select one of the values .01, .02, . . . , .99 at random for a proportion $A_1$. (Use a table of random numbers for this and later selections in this problem.) Now toss a coin: If the coin falls heads, take $A_2 = A_1$; if the coin falls tails, select at random for $A_2$ one of the 98 values different from $A_1$. Now select random samples of 100 from populations with proportions $A_1$ and $A_2$, calculate

$R$, and the posterior probability that $A_1 = A_2$. Would you have drawn a correct conclusion from your sample?

4. One hundred students are selected at random and watch their statistics lectures on television. One hundred similar students attend the same lectures in person. Of the first group, 60 score $A$ or $B$ on an exam, but only 50 of the second group score $A$ or $B$. Is this evidence for or against the hypothesis that television makes no difference in teaching statistics?

# ANSWERS

## ANSWERS: Set 1

1. (a) $\frac{1}{4}$  (b) $\frac{1}{2}$  (c) $\frac{1}{4}$
2. (a) $\frac{1}{6}$  (b) $\frac{1}{2}$  (c) $\frac{1}{3}$
3. $\frac{1}{2}, \frac{1}{4}, \frac{1}{8}, \frac{1}{16}, \frac{1}{16}$
4. (a) $\frac{12}{30}, \frac{8}{30}, \frac{8}{30}, \frac{2}{30}$  (b) $MM$  (c) 1
5. (a) $\frac{1}{2}$  (b) $\frac{1}{4}$  (c) $\frac{1}{4}$
   (d) $\frac{1}{6}$  (e) $\frac{1}{6}$  (f) $\frac{1}{12}$
6. $\frac{1}{2}, \frac{1}{3}, \frac{1}{6}, \frac{1}{6}, \frac{1}{3}, \frac{1}{2}$
7. (Partial)  yes, yes
8. $\frac{1}{6}, \frac{1}{4}$
9. $\frac{100}{271} = .37$
10. .9702
11. (a) $\frac{1}{32} = .031$  (b) .00128
12. .96
13. .93
14. $\frac{1}{2}, \frac{1}{4}, \frac{3}{4}$
15. $\frac{1}{4}$

## ANSWERS: Set 2

1. (*Partial*)

    $X$:  
    | $v$ | $p$ |
    |---|---|
    | 0 | $\frac{1}{3}$ |
    | 1 | $\frac{1}{3}$ |
    | 2 | $\frac{1}{3}$ |

    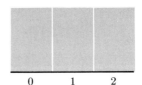

    $Y$: Same as $X$

    $X + Y$:
    | $v$ | $p$ |
    |---|---|
    | 1 | $\frac{1}{3}$ |
    | 2 | $\frac{1}{3}$ |
    | 3 | $\frac{1}{3}$ |

    $X - Y$:
    | $v$ | $p$ |
    |---|---|
    | $-2$ | $\frac{1}{6}$ |
    | $-1$ | $\frac{1}{3}$ |
    | 1 | $\frac{1}{3}$ |
    | 2 | $\frac{1}{6}$ |

    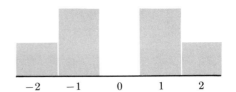

    $|X - Y|$:
    | $v$ | $p$ |
    |---|---|
    | 1 | $\frac{2}{3}$ |
    | 2 | $\frac{1}{3}$ |

    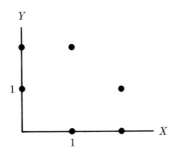

2. (*Partial*) $X$ and $Y$ same as in Problem 1.

    $X + Y$:
    | $v$ | $p$ |
    |---|---|
    | 0 | $\frac{1}{9}$ |
    | 1 | $\frac{2}{9}$ |
    | 2 | $\frac{3}{9}$ |
    | 3 | $\frac{2}{9}$ |
    | 4 | $\frac{1}{9}$ |

    $X - Y$:
    | $v$ | $p$ |
    |---|---|
    | $-2$ | $\frac{1}{9}$ |
    | $-1$ | $\frac{2}{9}$ |
    | 0 | $\frac{3}{9}$ |
    | 1 | $\frac{2}{9}$ |
    | 2 | $\frac{1}{9}$ |

**ANSWERS: SET 2**

$|X - Y|$:

| $v$ | $p$ |
|---|---|
| 0 | $\frac{3}{9}$ |
| 1 | $\frac{4}{9}$ |
| 2 | $\frac{2}{9}$ |

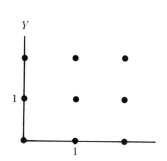

3. *(Partial)*

$X_1$:

| $v$ | $p$ |
|---|---|
| 0 | .2 |
| 1 | .8 |

$S_2$:

| $v$ | $p$ |
|---|---|
| 0 | .04 |
| 1 | .32 |
| 2 | .64 |

$S_3$:

| $v$ | $p$ |
|---|---|
| 0 | .008 |
| 1 | .096 |
| 2 | .384 |
| 3 | .512 |

$S_3 - S_2$: Same as $X_1$

$X_1 - X_2$:

| $v$ | $p$ |
|---|---|
| $-1$ | .16 |
| 0 | .68 |
| 1 | .16 |

4. *(Partial)*

| $v$ | $p$ | | | | |
|---|---|---|---|---|---|
| | .6 | .5 | | 0 | 1 |
| 0 | .064 | .125 | .729 | 1 | 0 |
| 1 | .288 | .375 | .243 | 0 | 0 |
| 2 | .432 | .375 | .027 | 0 | 0 |
| 3 | .216 | .125 | .001 | 0 | 1 |

5. (*Partial*)

| $X + 2$: | $v$ | $p$ |
|---|---|---|
| | 2 | .6 |
| | 4 | .2 |
| | 5 | .2 |

| $2X$: | $v$ | $p$ |
|---|---|---|
| | 0 | .6 |
| | 4 | .2 |
| | 6 | .2 |

| $X - 1$: | $v$ | $p$ |
|---|---|---|
| | −1 | .6 |
| | 1 | .2 |
| | 2 | .2 |

| $X^2$: | $v$ | $p$ |
|---|---|---|
| | 0 | .6 |
| | 4 | .2 |
| | 9 | .2 |

| $(X - 1)^2$: | $v$ | $p$ |
|---|---|---|
| | 1 | .8 |
| | 4 | .2 |

| $\dfrac{X}{X + 1}$: | $v$ | $p$ |
|---|---|---|
| | 0 | .6 |
| | $\frac{2}{3}$ | .2 |
| | $\frac{3}{4}$ | .2 |

Histogram for $\dfrac{X}{X + 1}$:

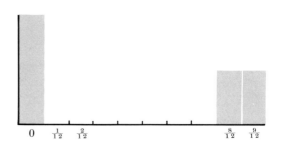

6. (*Partial*)

(a)

| $X$: | $v$ | $p$ |
|---|---|---|
| | 0 | $\frac{1}{9}$ |
| | 1 | $\frac{2}{9}$ |
| | 2 | $\frac{3}{9}$ |
| | 3 | $\frac{2}{9}$ |
| | 4 | $\frac{1}{9}$ |

| $Y$: | $v$ | $p$ |
|---|---|---|
| | 0 | $\frac{5}{9}$ |
| | 1 | $\frac{3}{9}$ |
| | 2 | $\frac{1}{9}$ |

| $X + Y$: | $v$ | $p$ |
|---|---|---|
| | 0 | $\frac{1}{9}$ |
| | 1 | $\frac{1}{9}$ |
| | 2 | $\frac{2}{9}$ |
| | 3 | $\frac{2}{9}$ |
| | 4 | $\frac{3}{9}$ |

(b)

| $Y - X$: | $v$ | $p$ |
|---|---|---|
| | 0 | 1 |

$Y - X$

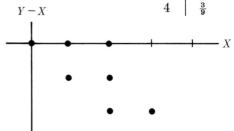

**ANSWERS: SET 2**

7. (Partial) $\frac{1}{16}, \frac{4}{16}, \frac{9}{16}$

| $X_2$: | $v$ | $p$ |
|---|---|---|
| | 0 | $\frac{1}{16}$ |
| | 1 | $\frac{3}{16}$ |
| | 2 | $\frac{5}{16}$ |
| | 3 | $\frac{7}{16}$ |

| $X_3$: | $v$ | $p$ |
|---|---|---|
| | 0 | $\frac{1}{64}$ |
| | 1 | $\frac{7}{64}$ |
| | 2 | $\frac{19}{64}$ |
| | 3 | $\frac{37}{64}$ |

$$P(X_n \leq k) = \left(\frac{k+1}{4}\right)^n \qquad k = 0, 1, 2, 3$$

8. (a)

| Outcome | $p$ |
|---|---|
| 00 | $\frac{6}{20}$ |
| 01 | $\frac{3}{20}$ |
| 02 | $\frac{3}{20}$ |
| 10 | $\frac{3}{20}$ |
| 12 | $\frac{1}{20}$ |
| 20 | $\frac{3}{20}$ |
| 21 | $\frac{1}{20}$ |

(b)

| Outcome | $p$ |
|---|---|
| 000 | $\frac{2}{20}$ |
| 001 | $\frac{2}{20}$ |
| 002 | $\frac{2}{20}$ |
| 010 | $\frac{2}{20}$ |
| 012 | $\frac{1}{20}$ |
| 020 | $\frac{2}{20}$ |
| 021 | $\frac{1}{20}$ |
| 100 | $\frac{2}{20}$ |
| 102 | $\frac{1}{20}$ |
| 120 | $\frac{1}{20}$ |
| 200 | $\frac{2}{20}$ |
| 201 | $\frac{1}{20}$ |
| 210 | $\frac{1}{20}$ |

(c)

| $v$ | $p$ |
|---|---|
| 0 | .3 |
| 1 | .3 |
| 2 | .3 |
| 3 | .1 |

(d)

| $v$ | $p$ |
|---|---|
| 0 | .3 |
| 1 | .6 |
| 4 | .1 |

9.

| $t$ | $P(X + Y \leq t)$ |
|---|---|
| 0 | 0 |
| .5 | $\frac{1}{8}$ |
| 1 | $\frac{1}{2}$ |
| 1.5 | $\frac{7}{8}$ |
| 2 | 1 |

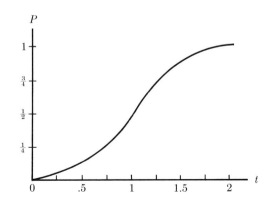

**BASIC STATISTICS**

10. (*Partial*)

    | $X_4$: | v | p |
    |---|---|---|
    | | 0 | $\frac{2}{9}$ |
    | | 2 | $\frac{7}{9}$ |

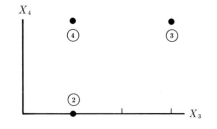

11. 
    | v | p |
    |---|---|
    | 2 | .16 |
    | 3 | .16 |
    | 4 | .36 |
    | 5 | .16 |
    | 6 | .16 |

## ANSWERS: Set 3

1. (*a*)  .1   (*b*)  .6   (*c*)  .3

    | v | p |
    |---|---|
    | 1 | .3 |
    | 3 | .4 |
    | 5 | .3 |

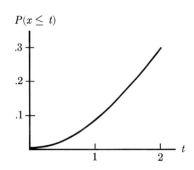
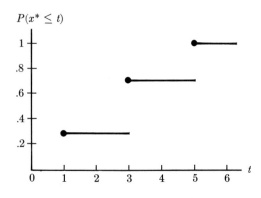

**ANSWERS: SET 3**

2.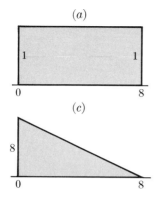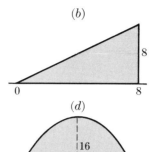

| $v$ | $p(a)$ | $p(b)$ | $p(c)$ | $p(d)$ |
|---|---|---|---|---|
| 1 | .25 | $\frac{1}{16}$ | $\frac{7}{16}$ | .15 |
| 3 | .25 | $\frac{3}{16}$ | $\frac{5}{16}$ | .35 |
| 5 | .25 | $\frac{5}{16}$ | $\frac{3}{16}$ | .35 |
| 7 | .25 | $\frac{7}{16}$ | $\frac{1}{16}$ | .15 |

3.  (*Partial*)

| $v$ | $p$ |
|---|---|
| 1 | .16 |
| 3 | .34 |
| 5 | .34 |
| 7 | .16 |

4.  (*Partial*) $P(2 \leq X \leq 4) = \frac{1}{3}$

| $v$ | $p$ |
|---|---|
| 1 | $\frac{1}{3}$ |
| 3 | $\frac{1}{3}$ |
| 5 | $\frac{1}{3}$ |

The density of $X$ is uniform on $0 \leq t \leq 6$.

5.  (*Partial*) $P(2 \leq X \leq 4) = \frac{1}{3}$

| $v$ | $p$ |
|---|---|
| 1 | $\frac{1}{9}$ |
| 3 | $\frac{3}{9}$ |
| 5 | $\frac{5}{9}$ |

The density of $X$ is $f(t) = \begin{cases} t & 0 \leq t \leq 6 \\ 0 & \text{elsewhere} \end{cases}$

## ANSWERS: Set 4

1. (*Partial*)

   | $v$ | $p(a)$ | $p(b)$ |
   |---|---|---|
   | 0 | $\frac{9}{36}$ | $\frac{3}{15}$ |
   | 1 | $\frac{12}{36}$ | $\frac{6}{15}$ |
   | 2 | $\frac{10}{36}$ | $\frac{4}{15}$ |
   | 3 | $\frac{4}{36}$ | $\frac{2}{15}$ |
   | 4 | $\frac{1}{36}$ | 0 |

   Mean income is $\frac{4}{3}$ in both cases. Replace 2, 4, not 0, giving mean income $\frac{7}{5}$.

2. $E(X) = 4$, $E(Y) = 2.5$, $E(X + Y) = 6.5$

3. (*Partial*)

   | $t$ | $E(X - t)^2$ |
   |---|---|
   | 0 | 2 |
   | 1 | 1 |
   | 2 | 2 |
   | 3 | 5 |

   $t = 1$   $E(X) = 1$

4. $E(X) = 500.5$, sum is 500, 500
5. (*Partial*) (a) $t$   (b) $t$   (c) $t^2$   (d) $t - t^2$
   $0 \leq E(X) \leq 1$,   $0 \leq E(X^2) - [E(X)]^2 \leq \frac{1}{4}$
6. (*Partial*) It can be done. $1 \leq E(X^2) \leq 2$
7. $\frac{7}{4}, \frac{3}{2}, \frac{15}{8}, \frac{31}{16}$, nearly 2   8. 2.5
9. (*Partial*) The exact values are: (a) 5, $\frac{100}{3}$   (b) 4, 24
10. (*Partial*) The exact values are: 3.5   15

11. (a)

    | $v$ | | $p$ | |
    |---|---|---|---|
    |  | .4 | .5 | .6 |
    | 3 | .16 | .25 | .36 |
    | 2 | .24 | .25 | .24 |
    | 1 | .24 | .25 | .24 |
    | $-6$ | .36 | .25 | .16 |
    | mean | $-.96$ | 0 | .84 |

    (b)

    |  | (1; 2,3) | (1; 0,2) | (1; 0,1) |
    |---|---|---|---|
    | $p = .4$ | $-.72$ | $-.44$ | $-.32$ |
    | $p = .6$ | .68 | .36 | .28 |

    (c) $E(W)$ has the same sign as $p - .5$.

ANSWERS: SET 5

12. .6, .6, 3 with or without replacement    13. 1, 2
14. .75  The exact value of $E(X)$ is $\frac{2}{3}$.
15. (a) 1   (b) $\frac{2}{3}$
16. (a) $\frac{13}{9}$   (b) $\frac{5}{3}$
17. (a) $\frac{25}{16}$   (b) 1

## ANSWERS: Set 5

1. (Partial)

| | $E(X)$ | $E(Y)$ | $\sigma(X)$ | $\sigma(Y)$ | $\sigma(X+Y)$ |
|---|---|---|---|---|---|
| (a) | $\frac{1}{2}$ | 1 | $\frac{1}{2}$ | .82 | .96 |
| (b) | $\frac{2}{3}$ | $\frac{4}{3}$ | .94 | .47 | .82 |
| (c) | $\frac{2}{3}$ | $\frac{2}{3}$ | .75 | .75 | .75 |
| (d) | 1 | $\frac{6}{5}$ | .89 | .75 | .98 |
| (e) | 1 | 1 | 1 | 1 | 1.4 |
| (f) | 0 | 1 | 0 | .82 | .82 |
| (g) | 1 | 1 | .82 | .82 | 0 |
| (h) | $\frac{4}{3}$ | $\frac{4}{3}$ | .47 | .94 | 1.2 |
| (i) | 1 | $\frac{4}{3}$ | .82 | .47 | .47 |

2. 4.5, 2.9    3. (a) 1.5, .87   (b) 1.5, .5
4. 68, 19, 64, 2, 72, 4, 16

5. (Partial)

| $n$ | $E(S_n)$ | $\sigma^2(S_n)$ | $E(Y_n)$ | $\sigma^2(Y_n)$ |
|---|---|---|---|---|
| 1 | .6 | .24 | .6 | .24 |
| 2 | 1.2 | .48 | .6 | .12 |
| 3 | 1.8 | .72 | .6 | .08 |
| 4 | 2.4 | .96 | .6 | .06 |

8. (Partial)

| | Largest $\sigma$ | $t$ value | Smallest $\sigma$ | $t$ value |
|---|---|---|---|---|
| (a) | .5 | .5 | 0 | 0 or 1 |
| (b) | 2.5 | .5 | 0 | 0 or 1 |
| (c) | .77 | .3 | .49 | 0 or .6 |
| (d) | .98 | .6 | .49 | 0 |
| (e) | .94 | near 0 or 2 | .82 | 1 |
| (f) | as large as you please | large | .82 | 1 |
| (g) | .78 | any | .78 | any |
| (h) | 1.6 | 2 | .78 | 1 |

9. (*Partial*)

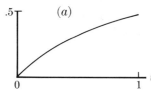

| | (a) Exact values: | (b) Exact values: |
|---|---|---|
| | $m = .63$ | $m = .5$ |
| 10. 1.4 | $\sigma = .25$ | $\sigma = .39$ |
| 11. .83 | $P = .61$ | $P = .46$ |

## ANSWERS: Set 6

1. (*Partial*)
   (a) mse:   2.5,   1.75,   1.5,   1.75,   2,   2.5
       Worth:  $-\frac{2}{3}$,   $-\frac{1}{6}$,   0,   $-\frac{1}{6}$,   $-\frac{2}{3}$
   (b) mse:   1.5,   3.5,   2.5,   1.42
       Worth:  0,   $-\frac{4}{3}$,   $-\frac{2}{3}$,   .053
   (c) 0, 1.5, 1 for $X = 0, 1, 2$ respectively
       Worth .25
   (e) .5, 2, 1 for $Y = 0, 1, 3$ respectively
       Worth .75,   no

2.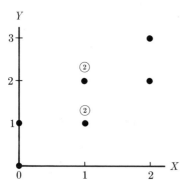

   (a) .5, 1.5, 2.5 for $X = 0, 1, 2$ respectively
       Worth $\frac{2}{3}$,   yes
   (b) 0, $\frac{2}{3}$, $\frac{4}{3}$, 2 for $Y = 0, 1, 2, 3$ respectively
       Worth $\frac{2}{3}$,   yes

3.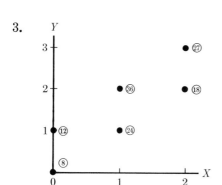

(a) .6, 1.6, 2.6 for $X = 0, 1, 2$ respectively
Worth $\frac{2}{3}$, yes
(b) Same as 2(b).
4. (*Partial*) Worths: (a) 0, 0   (b) $\frac{1}{3}$, 0

## ANSWERS: Set 7

1.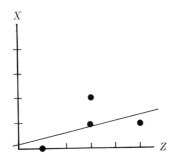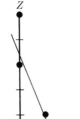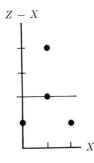

(a) $\rho^2 = .25$

$$U = \frac{Z+1}{4}$$

(b) $\rho^2 = \frac{2}{3}$

$$U = \frac{11 - 8Y}{3}$$

(c) $\rho^2 = 0$

$$U = 2$$

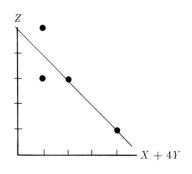
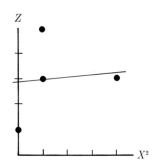

(d)  $\rho^2 = \frac{3}{4}$            (e)  $\rho^2 = \frac{1}{27}$

$$U = 5 - (X + 4Y) \qquad\qquad U = \frac{2X^2 + 24}{9}$$

4. For the second diagram, it is not possible to make $\rho^2 = 1$, but there are many ways to make $\rho^2$ nearly 1; for instance $(u,u)$ with $u$ large. There are many solutions to each of the others. For instance: first diagram (a) $(0, \frac{11}{3})$, (b) $(4,4)$; second diagram (a) $(1,2)$.

5. (*Partial*)

| a | | | | | b | | | | | $\rho^2$ | | | |
|---|---|---|---|---|---|---|---|---|---|---|---|---|---|
| h \ k | 1 | 3 | 10 | | h \ k | 1 | 3 | 10 | | h \ k | 1 | 3 | 10 |
| 1 | $\frac{1}{4}$ | $\frac{3}{4}$ | $\frac{20}{8}$ | | 1 | $\frac{1}{4}$ | $\frac{3}{4}$ | $\frac{5}{2}$ | | 1 | $\frac{1}{4}$ | $\frac{1}{4}$ | $\frac{1}{4}$ |
| 3 | $\frac{1}{4}$ | $\frac{3}{4}$ | $\frac{20}{8}$ | | 3 | $\frac{1}{12}$ | $\frac{1}{4}$ | $\frac{5}{6}$ | | 3 | $\frac{3}{4}$ | $\frac{3}{4}$ | $\frac{3}{4}$ |
| 10 | $\frac{10}{103}$ | $\frac{30}{103}$ | $\frac{100}{103}$ | | 10 | $\frac{1}{103}$ | $\frac{10}{103}$ | $\frac{100}{103}$ | | 10 | $\frac{100}{103}$ | $\frac{100}{103}$ | $\frac{100}{103}$ |

6. $(90G + 11)/140$, $\frac{81}{84}$      7. $(X + 3)/6$, $\frac{1}{6}$      8. .6

ANSWERS: Set 8

1. $(3X + 1)/2$, $(2X + Y)/2$, $X + 1$, .9, $\frac{49}{60}$, .95, .5, $\frac{8}{11}$, .71
3. (*Partial*)   $c = \dfrac{\text{cov}(X,Z)}{\text{cov}(X,X)}$    $d = \dfrac{\text{cov}(Y,Z)}{\text{cov}(Y,Y)}$
4. (a) $X_2 + .6$    (b) $(X_1 + X_3)/2$    (c) $X_2/2$
5. $X^2 - 2X + 1$

## ANSWERS: Set 9

1. $\frac{13}{12}$, $\frac{95}{144}$, .81

2. 
| $v$ | $p$ |
|---|---|
| 2 | .16 |
| 3 | .40 |
| 4 | .25 |
| 5 | .08 |
| 6 | .10 |
| 8 | .01 |

   Yes, from

| $v$ | $p$ |
|---|---|
| 1 | .4 |
| 2 | .5 |
| 4 | .1 |

5. From

| $v$ | $p$ |
|---|---|
| 0 | .5 |
| 1 | .5 |

6. $(.3)^6 = .000729$

| $v$ | $p$ |
|---|---|
| 0 | $1 - (.3)^6$ |
| 1 | $(.3)^6$ |

7. .4, $\frac{2}{3}$, .16, $\frac{2}{3}$, $\frac{2}{3}$, $(.4)^5 = .01024$, .98976

8. (*Partial*) $\frac{1}{36}$, $\frac{1}{3}$, it does work.

## ANSWERS: Set 10

1. .36, .48, .16, .22, .13, .38, .25, .15, .41, .41, 0, 0

2. (*Partial*)

$n = 5$, $p = .4$, $m = 2$, $\sigma = 1.1$, $P = .64$

3. (*Partial*) Graph $4p(1-p)^3$, .25, .42
4. 4 or 5, .35
5. (*Partial*) Graph $10p$, $10p(1-p)$, $p$, $p(1-p)/10$
6. (*Partial*) Graph $.4r$, $.24r$, $.4$, $.24/r$
7. (a) 1, 1, 1, 1, 1; 0, $\frac{1}{2}$, $\frac{2}{3}$, $\frac{4}{5}$, $\frac{9}{10}$ (b) 1,0 all n
   (c) 0, .25, .30, .33, $(.9)^{10} = .35$
8. (*Partial*) $r = \dfrac{(.3)(63)}{(.7)(38)} = \dfrac{189}{266}$ 30 is the most probable value.

10. (*Partial*)

| $k$ | $p(k)$ | $p^2(k)$ | $kp^2(k)$ | $(k+\frac{1}{2})p^2(k)$ |
|---|---|---|---|---|
| 1 | .5000 | .2500 | .2500 | .3750 |
| 3 | .3125 | .09766 | .2930 | .3418 |
| 5 | .2461 | .06057 | .3028 | .3331 |

For large $k$, $kp^2(k)$ is about $1/\pi = .3183$; that is, $p(k)$ is about $1/\sqrt{\pi k} = .5642/\sqrt{k}$.

11.

| $v$ | $p(a)$ | $p(b)$ | $p(c)$ |
|---|---|---|---|
| 0 | .32 | .16 | .4 |
| 1 | .56 | .48 | .4 |
| 2 | .12 | .36 | .2 |
| $E(S)$ | .8 | .8 | .8 |
| $\sigma^2(S)$ | .40 | .48 | .56 |

## ANSWERS: Set 11

1. .159, .774, .001, 675
3. .009 (exact $P = .011$)
4. .00016, .0009

5. (*Partial*) $X^{*2}$:

| $v$ | $p$ | |
|---|---|---|
|  | a | b |
| .09 | .452 | .458 |
| .81 | .318 | .319 |
| 2.25 | .158 | .156 |
| 4.41 | .056 | .053 |
| 7.29 | .014 | .012 |

**ANSWERS: SET 11**

6. .976

7. $\frac{63}{256} = .246$, .247

8.
9.

| $n$ | $p$ | $d$ | $P(|\bar{X} - p| \le d)$ |
|---|---|---|---|
| 225 | .5 | .02 | .45 |
| 225 | .5 | .05 | .87 |
| 225 | .7 | .02 | .49 |
| 225 | .7 | .05 | .90 |
| 900 | .5 | .02 | .77 |
| 900 | .5 | .05 | .997 |
| 900 | .7 | .02 | .81 |
| 900 | .7 | .05 | .999 |

10. .0399, $\sqrt{.001601}$ $(<.0410)$, $\sqrt{.001514}$ $(>.038)$

11. (*Partial*) (b)

| $v$ | $p$ |
|---|---|
| 0 | $\frac{1}{16}$ |
| 1 | $\frac{2}{16}$ |
| 2 | $\frac{3}{16}$ |
| 3 | $\frac{4}{16}$ |
| 4 | $\frac{3}{16}$ |
| 5 | $\frac{2}{16}$ |
| 6 | $\frac{1}{16}$ |

(c)

| $v$ | $p$ | |
|---|---|---|
| | *Exact* | *Approximate* |
| 0,12 | $a = .0039$ | .006 |
| 1,11 | $4a = .0156$ | .016 |
| 2,10 | $10a = .039$ | .037 |
| 3,9 | $20a = .078$ | .073 |
| 4,8 | $31a = .121$ | .120 |
| 5,7 | $40a = .156$ | .160 |
| 6 | $44a = .172$ | .176 |
| | where $a = \frac{1}{256}$ | |

**BASIC STATISTICS**            **136**

## ANSWERS: Set 12

1. $b$ if $B$, $a$ if $W$;   .75
3. (a)   $b$ if $B$, either if $W$;   .8      (b)   $b$ always,   .9

4.

| $\lambda$ | $A$ | $X$ | | | | .332,   .0294 |
|---|---|---|---|---|---|---|
| | | $GG$ | $GU$ | $UG$ | $UU$ | |
| $\frac{1}{3}$ | .1 | .01 | .09 | .09 | .81 | |
| $\frac{1}{3}$ | .2 | .04 | .16 | .16 | .64 | |
| $\frac{1}{3}$ | .5 | .25 | .25 | .25 | .25 | |

5. (*Partial*)   .332,   .0294
6. (*Partial*)   (a)   2 if $X = 0$, either if $X = 1$, 1 if $X = 2$;   $\frac{2}{3}$
               (b)   Same decision method as (a);   $\frac{5}{6}$

7. (*Partial*)    **DATA TABLE**

| $\lambda$ | $A$ | $X$ | | | | Overall chance $= \frac{7}{12}$ |
|---|---|---|---|---|---|---|
| | | T | H | I | S | |
| $\frac{1}{3}$ | THIS | .25 | .25 | .25 | .25 | |
| $\frac{1}{3}$ | IS | 0 | 0 | .5 | .5 | |
| $\frac{1}{3}$ | IT | .5 | 0 | .5 | 0 | |

8. (*Partial*)    **DATA TABLE**

| $\lambda$ | $Y$ | $X$ | | | | Overall mse $= .9$ |
|---|---|---|---|---|---|---|
| | | 0 | 1 | 2 | 3 | |
| .4 | 0 | $\frac{1}{4}$ | $\frac{1}{4}$ | $\frac{1}{4}$ | $\frac{1}{4}$ | |
| .2 | 1 | $\frac{1}{2}$ | $\frac{1}{2}$ | 0 | 0 | |
| .3 | 2 | $\frac{1}{3}$ | $\frac{1}{2}$ | $\frac{1}{3}$ | 0 | |
| .1 | 3 | 1 | 0 | 0 | 0 | |

**ANSWERS: SET 13**

9.

| $n$ | $p(1)$ | $p(2)$ | $p(3)$ | $p(4)$ | $p(5)$ |
|---|---|---|---|---|---|
| 3 | $\frac{20}{47}$ | $\frac{20}{47}$ | $\frac{20}{47}$ | 0 | 0 |
| 4 | $\frac{15}{47}$ | $\frac{15}{47}$ | $\frac{15}{47}$ | $\frac{5}{9}$ | 0 |
| 5 | $\frac{12}{47}$ | $\frac{12}{47}$ | $\frac{12}{47}$ | $\frac{4}{9}$ | 1 |

**ANSWERS: Set 13**

1.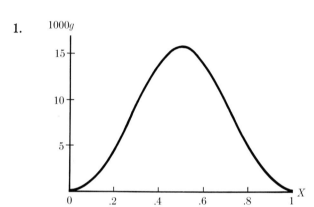

Exact $\sigma^2$ is $\frac{1}{36}$.

4. $P_1 = .0256 \qquad P_2 = .3840 \qquad P_3 = .5904 \qquad E(A^*) = .76688$

5.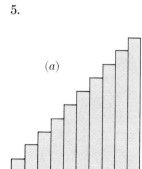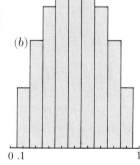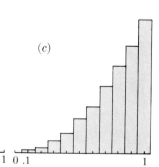

**BASIC STATISTICS**

ANSWERS: Set 14

3. (*Partial*)  .3024

ANSWERS: Set 16

1. (*Partial*)   2.02,  .156,  2.1;    32,  nearly 0,  .2
4. For

# APPENDIX

**TABLE OF SQUARES, 10–99**

|   | 0 | 1 | 2 | 3 | 4 | 5 | 6 | 7 | 8 | 9 |
|---|------|------|------|------|------|------|------|------|------|------|
| 1 | 100  | 121  | 144  | 169  | 196  | 225  | 256  | 289  | 324  | 361  |
| 2 | 400  | 441  | 484  | 529  | 576  | 625  | 676  | 729  | 784  | 841  |
| 3 | 900  | 961  | 1024 | 1089 | 1156 | 1225 | 1296 | 1369 | 1444 | 1521 |
| 4 | 1600 | 1681 | 1764 | 1849 | 1936 | 2025 | 2116 | 2209 | 2304 | 2401 |
| 5 | 2500 | 2601 | 2704 | 2809 | 2916 | 3025 | 3136 | 3249 | 3364 | 3481 |
| 6 | 3600 | 3721 | 3844 | 3969 | 4096 | 4225 | 4356 | 4489 | 4624 | 4761 |
| 7 | 4900 | 5041 | 5184 | 5329 | 5476 | 5625 | 5776 | 5929 | 6084 | 6241 |
| 8 | 6400 | 6561 | 6724 | 6889 | 7056 | 7225 | 7396 | 7569 | 7744 | 7921 |
| 9 | 8100 | 8281 | 8464 | 8649 | 8836 | 9025 | 9216 | 9409 | 9604 | 9801 |

**BASIC STATISTICS**

**NORMAL DISTRIBUTION TABLE**

| $t$ | $h(t)$ | $H(t)$ | $t$ | $h(t)$ | $H(t)$ | $t$ | $h(t)$ | $H(t)$ |
|---|---|---|---|---|---|---|---|---|
| .00 | .399 | .500 | 1.50 | .130 | .0668 | 3.00 | .0$^2$443 | .0$^2$135 |
| .05 | .398 | .480 | 1.55 | .120 | .0606 | 3.05 | .0$^2$381 | .0$^2$115 |
| .10 | .397 | .460 | 1.60 | .111 | .0548 | 3.10 | .0$^2$327 | .0$^3$968 |
| .15 | .394 | .440 | 1.65 | .102 | .0495 | 3.15 | .0$^2$279 | .0$^3$816 |
| .20 | .391 | .421 | 1.70 | .0940 | .0446 | 3.20 | .0$^2$238 | .0$^3$687 |
| .25 | .387 | .401 | 1.75 | .0863 | .0401 | 3.25 | .0$^2$203 | .0$^3$577 |
| .30 | .381 | .382 | 1.80 | .0790 | .0359 | 3.30 | .0$^2$172 | .0$^3$483 |
| .35 | .375 | .363 | 1.85 | .0721 | .0322 | 3.35 | .0$^2$146 | .0$^3$404 |
| .40 | .368 | .345 | 1.90 | .0656 | .0287 | 3.40 | .0$^2$123 | .0$^3$337 |
| .45 | .361 | .326 | 1.95 | .0596 | .0256 | 3.45 | .0$^2$104 | .0$^3$280 |
| .50 | .352 | .309 | 2.00 | .0540 | .0228 | 3.50 | .0$^3$873 | .0$^3$233 |
| .55 | .343 | .291 | 2.05 | .0488 | .0202 | 3.55 | .0$^3$732 | .0$^3$193 |
| .60 | .333 | .274 | 2.10 | .0440 | .0179 | 3.60 | .0$^3$612 | .0$^3$159 |
| .65 | .323 | .258 | 2.15 | .0396 | .0158 | 3.65 | .0$^3$510 | .0$^3$131 |
| .70 | .312 | .242 | 2.20 | .0355 | .0139 | 3.70 | .0$^3$425 | .0$^3$108 |
| .75 | .301 | .227 | 2.25 | .0317 | .0122 | 3.75 | .0$^3$353 | .0$^4$884 |
| .80 | .290 | .212 | 2.30 | .0283 | .0107 | 3.80 | .0$^3$292 | .0$^4$723 |
| .85 | .278 | .198 | 2.35 | .0252 | .0$^2$939 | 3.85 | .0$^3$241 | .0$^4$591 |
| .90 | .266 | .184 | 2.40 | .0224 | .0$^2$820 | 3.90 | .0$^3$199 | .0$^4$481 |
| .95 | .254 | .171 | 2.45 | .0198 | .0$^2$714 | 3.95 | .0$^3$163 | .0$^4$391 |
| 1.00 | .242 | .159 | 2.50 | .0175 | .0$^2$621 | 4.00 | .0$^3$134 | .0$^4$317 |
| 1.05 | .230 | .147 | 2.55 | .0154 | .0$^2$539 | 4.05 | .0$^3$109 | .0$^4$256 |
| 1.10 | .218 | .136 | 2.60 | .0136 | .0$^2$466 | 4.10 | .0$^4$893 | .0$^4$207 |
| 1.15 | .206 | .125 | 2.65 | .0119 | .0$^2$402 | 4.15 | .0$^4$726 | .0$^4$166 |
| 1.20 | .194 | .115 | 2.70 | .0104 | .0$^2$347 | 4.20 | .0$^4$589 | .0$^4$133 |
| 1.25 | .183 | .106 | 2.75 | .0$^2$909 | .0$^2$298 | 4.25 | .0$^4$477 | .0$^4$107 |
| 1.30 | .171 | .0968 | 2.80 | .0$^2$792 | .0$^2$256 | 4.30 | .0$^4$385 | .0$^5$854 |
| 1.35 | .160 | .0885 | 2.85 | .0$^2$687 | .0$^2$219 | 4.35 | .0$^4$310 | .0$^5$681 |
| 1.40 | .150 | .0808 | 2.90 | .0$^2$595 | .0$^2$187 | 4.40 | .0$^4$249 | .0$^5$541 |
| 1.45 | .139 | .0735 | 2.95 | .0$^2$514 | .0$^2$159 | 4.45 | .0$^4$200 | .0$^5$429 |

# INDEX

Allocation, 111
Approximating variable, 22, 31
Approximation, 22

Best linear predictor, 52ff, 60
   mean of, 54
   worth of, 52, 62
Best predictor, 51
Binomial distribution, 74ff
Binomial parameters, 75
Binomial variable, 75
   mean and variance of, 75

$C(n,k)$, 73, 76
   table of, 73
Chi-square statistic, 115
   level of, 115
Conditional probability, 4
Correlation, 51ff, 59
Correlation coefficient, 52
Covariance, 52
   of independent variables, 70

Data table, 89
Density, 20, 21

Difference between two proportions, 109
Dispersion, 40
Distribution, 12
    for approximating variable, 22ff
    binomial, 74ff
Drawing with replacement, 5
Drawing without replacement, 5

Event, 2
    certain, 2
    impossible, 2
Events, mutually exclusive, 2
Expected $2 \times 2$ table, 115
Expected value (mean), 26

Function, linear, 52

Histogram, 13, 20
    for approximating variable, 22

Inference, 88ff

Joint distribution table, 90

Least-squares line (*see* Least-squares linear predictor)
Least-squares linear predictor, 54
    drawing of, 54
    slope of, 55
Level of chi square, 115
Linear function, 52

Marginal (overall) distribution, 90
Mean (expected value), 26
    of binomial variable, 75
    as predictor, 48

Mean (expected value), properties of, 29
    of random sample, 67
    of sample mean, 67
Mean squared error (mse), 37, 92ff
Model for an experiment, 89
Multiplication rule, 4
Mutually exclusive, 2

Normal approximation, 82ff
Normal curve, 81
Normal distribution, table of, 140

Outcome, 89
Overall (marginal) distribution, 90

Parameter, 75, 89
    binomial, 75
Partial correlation coefficient, 60
Polya urn scheme, 6
Posterior density, 97
Posterior distribution, 91
Posterior distribution table, 90, 91
Predictor, 37
    best, 37, 46ff, 51
    worth of, 47
Prior density, 97
Prior distribution, 89
Probability, definition of, 1
Probability from density, 21
Proportions, 96ff
    independent, 108

Random number table, 70
Random sample, 66
    mean of, 67
Random sampling, 65
Random selection, 1

Sample mean, 67
Sample size, 74
Scatter diagram, 15
Selection at random, 1, 3
Slope, 55
Squared correlation coefficient, 52
Squared multiple correlation coefficient, 60
Squared partial correlation coefficient, 60
Squares of numbers, table of, 139
Standard deviation, 38, 39
   properties of, 39
   of sample mean, 67
Standard normal variable, 82
Systems, 29

Table of squares, 139

$(u,v)$ density, 102
   graph of, 103
   mean and variance of, 104

Variable, 12
   approximating, 22, 31
   binomial, 75
   distribution of, 12
   standard normal, 82
   value of, 12
Variables, independent, 64, 65
   sums, differences, products, and quotients of, 13, 14
Variance, 37ff
   of binomial variable, 75
   properties of, 39
   of sum of independent variables, 65

Weighted average between proportions, 109